T0214007

Text Mining for Information Professionals

Manika Lamba • Margam Madhusudhan

Text Mining for Information Professionals

An Uncharted Territory

 Springer

Manika Lamba
Library and Information Science
University of Delhi
Delhi, India

Margam Madhusudhan
Library and Information Science
University of Delhi
Delhi, India

ISBN 978-3-030-85084-5   ISBN 978-3-030-85085-2   (eBook)
https://doi.org/10.1007/978-3-030-85085-2

This Springer imprint is published by the registered company Springer Nature Switzerland AG.
The registered company address is: Gewerbestrasse 11, 6330 Cham, Switzerland

*To my grandparents and brother for their
endless support and constant encouragement
– Manika*

Preface

Machine learning and artificial intelligence are the futuristic approaches dominating all disciplines currently. Data analytics, data mining, and data science are some of the significant sub-domains with a strong market of research and jobs at the moment in the world. With a long leap from paper to digitization, the burden on libraries and librarians has increased to manage, organize, and generate knowledge from such a massive amount of data stored in their repositories/databases/websites. As libraries deal with a higher percentage of textual data daily, this book focuses primarily on textual data and presents various text mining approaches through a new lens. Text mining is a very efficient, fast, and effective way of managing and extracting knowledge from existing data stored in the archives of libraries. This book will make every library and information professional competent to use text mining in their daily life and get the best out of it by serving their patrons, researchers, faculty, or scientists with new services. Text mining techniques can be applied to any library type, be it a school, university, or special library by the librarians. It will help to provide the *right information* to the *right user* at the *right time* by providing services like recommendation services, current awareness services, or selective dissemination services to its users.

From understanding different types and forms of data to case studies covering primary research showing the application of each text mining approach on data retrieved from various resources, this book will be a must-read for all library professionals interested in text mining and its application in libraries. Additionally, this book will be helpful to archivists, digital curators, or any other humanities and social science professionals who want to understand the basic theory behind text data, text mining, and various tools and techniques available to solve and visualize their research problems.

Key points of the book

1. Contains 14 demonstrative step-by-step case studies which show how to conduct 8 different text mining and visualization approaches on 9 distinct data type sources
2. Provides case studies demonstrating the use of five open-source software for both non-programmers and programmers

3. Reproduces six case studies using R programming in the cloud without having to install any software
4. Story section presenting 17 real-life experiences of the application of text mining methods and tools by 24 librarians/researchers/faculty/publishers
5. Elucidates 19 open-source text mining and visualization tools with their advantages and disadvantages
6. Illustrates various use cases that show how text mining strategies have been used in different ways in libraries across the globe

The book contains *11 chapters*, *14 case studies* showing 8 different text mining and visualization approaches, and *17 stories*. A website (https://textmining-infopros.github.io/) and a GitHub account (https://github.com/textmining-infopros) are also maintained for the book. They contain the code, data, and notebooks for the case studies; a summary of all the stories shared by the librarians/faculty; and hyperlinks to open an interactive virtual RStudio/Jupyter Notebook environment. The interactive virtual environment runs case studies based on the R programming language for hands-on practice in the cloud without installing any software. Text mining is a topic of international interest, and this book has been written to meet the reading interests of both national and international audiences. It will be appropriate for both beginners and advanced-level readers as it has been written keeping their information needs in mind.

Many books in the market are written to meet the need of computer science professionals on text mining, whereas there are very few books on text mining for librarians. Also, the books present in the market on this topic are very difficult for non-programmers to understand. They may contain lots of jargon, which may not be easily understood by a library professional. In contrast, this book focuses on a basic theoretical framework dealing with the problems, solutions, and applications of text mining and its various facets in the form of *case studies*, *use cases*, and *stories*.

Delhi, India   Manika Lamba
Delhi, India   Margam Madhusudhan
June 2021

Acknowledgments

The authors are immensely grateful to Alex Wermer-Colan, Amy J. Kirchhoff, Anne M. Brown, C. Cozette Comer, Carady DeSimone, Chreston Miller, Cody Hennesy, Issac Williams, Jacob Lahne, Jonathan Briganti, Jordan Patterson, Karen Harker, Leah Hamilton, Lighton Phiri, Manisha Bolina, Manoj Kumar Verma, Marcella Fredriksson, Maya Deori, Michael J. Stamper, Nathaniel D. Porter, Parthasarathi Mukhopadhyay, Rachel Miles, Sephra Byrne, and Vinit Kumar for sharing their personal experience of using different text mining approaches for the *story section* of the book.

Contents

Chapter 1
The Computational Library

Abstract This chapter introduces the concept of computational research and thinking in libraries. It covers (i) the basic concept of text mining, (ii) the need for text mining in libraries, (iii) understanding text characteristics, and (iv) identifying different text mining problems, which include document classification/text categorization, information retrieval, clustering, and information extraction. It presents various use cases of libraries that have applied text mining techniques and enumerates various costs, limitations, and benefits related to text mining. This chapter is followed by a case study showing the clustering of documents using two different tools.

1.1 Computational Thinking

With the emergence of digital platforms, software, and media development, the exponential growth of data was inevitable. Computational methods are being used in many social science and humanities disciplines to exploit this digital trace data to answer different research questions. A *computational library* uses computational research in libraries to curate, steward, preserve, and provide access to the generated information. It is about making the library's collection accessible by applications, algorithms, machines, and "people to support data-intensive research from start to finish, from data licensing, acquisition, and management to long-term data storage and access" [1]. It also helps to enable "computational access to the full suite of information and data managed by the libraries" [1].

"With a long history of providing information for use in research, managing it equitably, advocating for an open scholarship, and taking the long view on durable information access, libraries are uniquely positioned to help build a computational research support model at scale" [1]. For instance, in 2019, MIT Libraries launched their discovery index and public API—TIMDEX.[1] TIMDEX indexes theses, online journals, and the entire catalog of electronic and print books subscribed, purchased,

---

[1] https://timdex.mit.edu/.

and owned by the MIT Libraries. A library can provide a *programmable environment* to its community by using computational techniques and methods. It can give its community the platform to design its code to access and analyze its collections. "Librarians know better than most people how information is collected, assembled, and organized, so they know where things can go wrong. Thus, they can help ensure scholars are aware of the shortcomings of the data they use and can help mitigate those impacts" [1].

Computational thinking corresponds to the thought process that is employed to develop problems and solutions [2]. It "includes breaking down problems into smaller parts, looking for patterns, identifying principles that generate these patterns, and developing instructions that the computers, machines, and people, can understand. It is an approach to critical thinking that can be used to solve problems across all disciplines" [3]. Librarians play a crucial role in information literacy and lifelong learning. They interact with patrons of all ages to assist with various "literacies, from text-based literacies to media literacy, financial literacy, and more. For instance, when libraries host *build a robot* activities, *design a website* programs, or enable patrons to create a video game, they gain computational thinking literacies" [3]. *Libraries Ready to Code (RtC)*[2] is an example of a computational literacy initiative created by Google's Computer Science Education[3] team and the American Library Association (ALA)[4] (Fig. 1.1). It is an online set of resources in partnership with 30 libraries to guide libraries to facilitate coding and computational thinking programs for youth and the library staff that support them.

Anne M. Brown, is an Assistant Professor and Science Informative Consultant at the Virginia Polytechnic Institute and State University, Virginia, and leads DataBridge in collaboration with **Jonathan Briganti**, DataBridge Manager.

Story: Leverage Undergraduate Student Researchers to Deliver Text Data Mining Services to Library Users
To expand the reach and meet the demand of consultation requests in Data Services at Virginia Tech (VT) University Libraries, DataBridge was formed to train students in the foundations of data science and apply those skills to real-world research. DataBridge aims to achieve two goals—(1) enhance the research and teaching mission of the University and Libraries and (2) have more human power available to complete detailed, embedded consultations that require specialized analysis and tools.

In the past year, DataBridge has completed two natural language processing (NLP) and text data mining (TDM) collaborations with VT researchers. The first project examined peer-reviewed articles about antimicrobial resis-

(continued)

---

[2] https://www.ala.org/tools/readytocode/.

[3] https://edu.google.com/intl/en_in/computer-science/.

[4] https://www.ala.org/.

tance (AMR) through a One Health lens. The aim was to identify shifts in vocabulary, word popularity over time, word usage based on major discipline domains, and cluster analysis. First, we used ContentMine "getpapers"[a], an open-source JavaScript package, to pull thousands of open-access articles from the European PubMed Collection (EuroPMC)[b]. We then used the Natural Language Toolkit (NLTK) and pandas in Python to convert articles into raw text, standardized words, and bigrams and to perform text mining. Finally, students processed and presented the data within interactive dashboards created through Tableau. This enabled the researchers to dive into the data, ask more in-depth questions about the methodology, see the scope of the field, and provide feedback throughout the process. Ultimately, we provided a specialized series of reports that the researchers are now using in a peer-reviewed manuscript about how AMR is discussed across multiple domains.

With the AMR project in our portfolio, more TDM- and NLP-based projects quickly followed. An active NLP- and TDM-based project aims to monitor the deluge of scientific articles being published on the coronavirus pandemic. Our students are using NLTK as our NLP foundation and integrating IBM Watson Tone Analyzer to ascertain the sentiment of these articles. The output of this project is continually evolving to match the needs of researchers pivoting to COVID-19 TDM analysis.

For both projects, DataBridge and the Libraries act as the technical subteam within the research groups we partner with, facilitating frequent communication with our workers and domain expert clients. Working with newly trained data scientists, as we primarily "employ" undergraduates, presents a challenge. With clearly defined documentation, training, validation, and communication protocols, this barrier is instead a great opportunity to strengthen training and increase the skill level of all student employees.

DataBridge has a public-facing Open Science Framework (OSF)[c] page and GitHub[d], containing all code created and used, replication steps, documentation, and CC-BY licenses. A link to live visualizations is stored within the OSF, and once the project has left active development, a ZIP file is sent to our data archival repository to preserve the code. By using open-source data and code libraries and choosing low-cost hosting strategies, we provide these resources as freely and easily accessible as possible. We are reaching the edge of more continued, in-depth TDM work with clients across campus and are excited to utilize student engagement for both the research outputs and educational opportunities.

[a] *https://github.com/ContentMine/getpapers*
[b] *https://europepmc.org/*
[c] *https://osf.io/82n73/*
[d] *https://github.com/databridgevt/AMR-Content-Mine*

(continued)

Cody Hennesy is a Journalism and Digital Media Librarian at the University of Minnesota, Twin Cities, where he works with computational researchers in the humanities and social sciences. He founded the Text as Data Practice Group with Wanda Marsolek and Alexis Logsdon.

Story—University of Minnesota: Text as Data Practice Group

Librarians at the University of Minnesota Twin Cities launched the Text as Data Practice Group[a] after observing an increasing number of graduate students, staff, and faculty who were hoping to learn more about computational text analysis methods. While some disciplines embed computational research into their curriculum in some form, many students in the social sciences and humanities still have few options to engage with computational text analysis methods in the classroom. Librarians and staff at Liberal Arts Technologies and Innovation Services (LATIS) have strived to fill that gap by collaboratively teaching introductory workshops on methods such as sentiment analysis and topic modeling, as well as web scraping and APIs, but it was difficult to develop enough content to keep up with growing demand. Students also show a wide range of preferences for different computational languages and packages, methods, and corpora, making it challenging to develop workshops to suit the broad scope of needs. The idea behind the practice group was to offer monthly peer-to-peer learning opportunities focusing on specific computational tools or methods using freely available online tutorials. This format allows the organizing team to provide a wide variety of learning opportunities without possessing deep expertise in each specific tool and method that is covered. The group is discipline-agnostic, welcomes all skill levels, and operates in the spirit of creating community. An early meeting brought together a mixture of students, staff, and faculty to work through the Programming Historian's tutorial on sentiment analysis[b] together.

Following the University's shutdown during the COVID-19 pandemic, the group pivoted to host online lightning talks from faculty, staff, and graduate students about their own research using text analysis methods. This switch significantly broadened the disciplinary reach of the practice group. Presentations have included sessions on detecting hate speech on Twitter, using sentiment analysis on Supreme Court documents, software development of text-based games, topic modeling of scholarly literature, and quantitative content analysis of news sources.

[a] *https://libguides.umn.edu/text-data-group*
[b] *https://programminghistorian.org/en/lessons/sentiment-analysis*

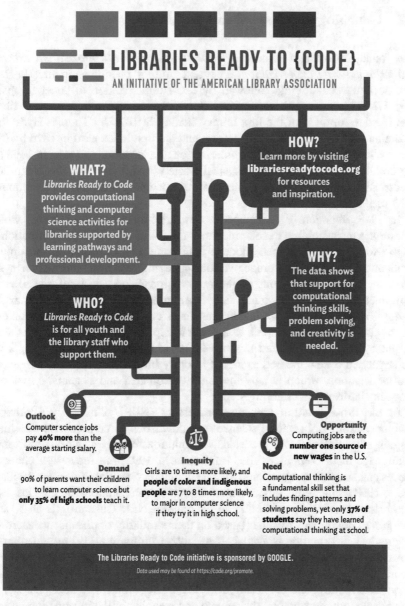

Fig. 1.1 Libraries Ready to Code (©2018 American Library Association, reprinted with permission by the American Library Association [4])

1.2   Genealogy of Text Mining in Libraries

The foundation of text mining is *information retrieval*, *information extraction*, and *information summarization*. The text mining phenomenon first began for text document cataloging, followed by text summarization to generate abstracts (Fig. 1.2). The earliest instance of text classification and summarization in libraries was the development of the first library catalog in 1674 by Thomas Hyde for the Bodleian Library, University of Oxford, and the first index card in 1876 by Melvil Dewey. It was followed by the summarization of a large body of texts in 1898 (from the collaboration between the Physical Society of London and the Institution of Electrical Engineers) and the generation of document abstracts by a computer at IBM in 1958 by Luhn [5].

In 1948, Shannon [7] developed a new area of information theory, which is among the most notable developments of the twentieth century. Information flow in the form of the Internet, modern data compression protocols, manipulating applications, document storage, various indexing systems, and search systems are some of the applications of the information theory. In 1950, the science of bibliometrics came into existence. It gives a numerical measure to analyze texts. It is an application of text processing that results in a collection of essential articles that can track the development path of a given discipline and is analogous to the word frequency calculation in text mining. In 1961, Doyle extended on Luhn's work. He suggested a new method to classify library information into word frequencies and associations, which is now the highly automated and systematic method for browsing information in libraries.

In the 1960s, natural language processing (NLP), a hybrid discipline, was developed from information science and linguistics to comprehend how natural languages are learned and modeled. The initial efforts of using NLP to translate a language on a computer failed, and soon in 1995, the focus was changed to processing answers to questions [8]. The computers' availability in the 1960s gave rise to NLP applications on computers known as computational linguistics. Luhn's [5] abstract generation method is an example of NLP. Clustering is an NLP task that groups documents together based on their similarity or distance measure when no previous information is available. In 1992, Cutting et al. [9] provided an early clustering analysis of browsing a document collection when a query could not be created. Subsequently, in 2002, Tombros et al. [10] used query-based clustering to perform hierarchical clustering of a document corpus. The next phase of NLP was related to understanding the context and meaning of the information instead of emphasizing the words used in the documents. Such developments were observed in the field of bibliometrics, where the context of documents was considered [11]. Modern text mining followed a similar path and arose from the above developments in NLP through the 1990s. In the 2000s, NLP practitioners can either use the domain-independent stemming and parsing strategies to build features or use the newer text categorization to tag documents.

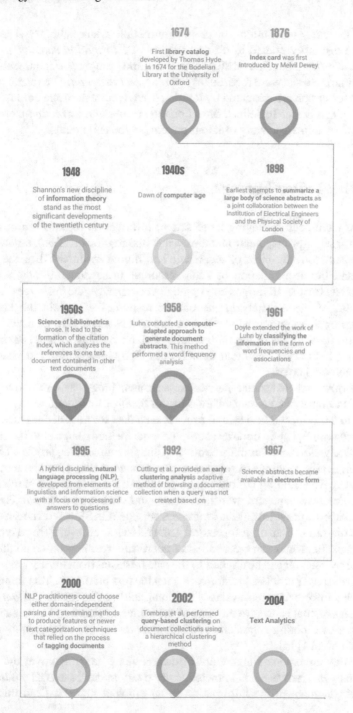

Fig. 1.2 History of text mining in libraries (Adapted from Miner et al. [6])

Thus, the early text mining tasks in information science, library science, and NLP were primarily related to different forms of *information summarization* and *information retrieval*, such as abstracts, indexes, and grouping of documents, but the later text mining tasks were focused on *information extraction*. The relationship and content information are extracted by tagging each document of the corpus. Modern text mining can be partly defined based on the *information extraction* methods that support information discovery of latent patterns in the text bodies.

1.3   What Is Text Mining?

Librarians have been managing large sets of information in the form of e-books, journal articles, e-prints, electronic theses and dissertations (ETDs), e-newspapers, and websites from the birth of electronic and digital resources. But the passage of time and the accumulation of such resources in several millions have made it impossible for the librarians to organize and manage the resources, let alone make sense of them. Analyzing electronic resources to unfold the knowledge encompassing them and provide efficient and timely services to its users is one of the vital responsibilities of the millennial librarians in today's digital environment. The application of text mining in the field of LIS by information professionals[5] is still new and uncharted.

"Platforms such as social networks, websites, blogs, and research journals generate an enormous amount of data in the form of unstructured text, and it is essential to analyze, synthesize, and process such data for knowledge discovery and decision-making" [12]. In libraries, computational methods, such as text mining, can be extensively used for searching, exploring, and recommending articles. Librarians can search similar readings to suggest to the users what they could be interested in reading next. Text mining not only helps the librarians in finding a similar type of journal articles, newspapers, websites, blogs, and other e-resources in text format but also helps to provide current awareness service (CAS), reference service, and selective dissemination of information (SDI) service efficiently in a very short period. Thus, librarians can create a list of interesting readings based on the queries submitted by users. It can be applied to the data retrieved from library blogs, journal articles, newspaper articles, literature, or social media platforms. Text mining is not only a quick information retrieval tool for librarians but also fulfills the *fourth law of library science*, that is, to save the time of the reader, as stated by Ranganathan [13]. Recently, a text mining approach called *topic modeling* has been gaining attention in the field of LIS [14].

Every day an exponential amount of data is being generated over the Internet. For several years, advances in knowledge discovery in databases (KDD) have been undertaken to manage the information in an efficient manner. Data mining is a

---

[5] In this book, information professionals and librarians are used interchangeably.

part of the KDD process, which identifies the hidden patterns in large information repositories. It involves several information extraction techniques, such as regression models, association rules, Bayesian methods, decision trees, neural networks, etc. "Text mining is a specialized interdisciplinary field combining techniques from linguistics, computer science, and statistics to build tools that can efficiently retrieve and extract information from the digital text" [15]. It assists in the automatic classification of documents.

"Text data is symmetrical content which contains knowledge. Text mining includes both text data and non-text data" [16]. Textual data is a form of unstructured data that humans can easily interpret and process but is difficult for the machines to perform the same. This difficulty has generated the need to use methods and algorithms to process the information. Thus, text mining is the process of extorting knowledge from a textual source, which can be structured as RDBMS data or semi-structured in an XML or JSON file and unstructured data as sources, for instance, textual documents, images, and videos. "In simple words, text mining is about looking for patterns in text. Depending on a study, text mining can be used to mine knowledge about a language; its usage and pattern across the text" [16].

"Generally, text mining and data mining are considered similar to one another, with a perception that the same techniques may be employed in both concepts to mine text. However, both are different in the sense that data mining involves structured data, while text deals with certain features and is relatively unstructured, and usually requires preprocessing (Table 1.1). Furthermore, text mining is an interrelated field with Natural Language Processing" [17].

"Text mining and text analytics are the terms usually used as same, but there is a minor difference between them. Text mining is more focused on the process of text mining, whereas text analytics is focused more on the results. In text mining, one wants to turn text data into high-quality information and actionable knowledge" [16]. As we know now, "text retrieval is a preprocessor for any text mining, that is, it turns big text data into a relatively small amount of most relevant text data and is needed to solve a particular problem with minimum human effort. It mainly consists of (i) NLP and text representation, (ii) word association mining and analysis, (iii) topic modeling and analysis, (iv) opinion mining and sentiment analysis, and (v) text-based predictions" [16]. In general, a text mining process has five steps, which are as follows (Fig. 1.3):

1. **Problem Identification**: Objectives of a research problem are identified;
2. **Pre-processing**: Data is being prepared for pattern extraction where the "raw text is transformed into some data representation format that can be used as input for the knowledge extraction algorithms" [19].
3. **Pattern Extraction**: "Suitable algorithm to extract the hidden patterns is applied. The algorithm is chosen based on the data available and the type of pattern that is expected" [19].
4. **Post-processing**: "Extracted knowledge is evaluated in this step" [19].

Table 1.1 Difference between data mining and text mining

Data mining	Text mining
It "is the core stage of the knowledge discovery process that aims to extract interesting and potentially useful information from data" [17]	It is the process of automatically extracting information from the text with the aim of generating new knowledge
It is at its maturity phase	It is an emerging area
Sample data is usually in numbers	Sample data is usually in text written by and for humans
They are usually structured in nature "usually sitting in a database and not necessarily generated by humans (it could be generated by a research instrument)" [18]	They are usually unstructured in nature
They are easy to analyze and process	They are difficult to analyze and process in comparison to data mining
It is easily understood by both machine and human	It is difficult to be understood by the machine
Usually, the statistical approach is considered to analyze the data	Clustering, classification, association, topic modeling, prediction analysis, visualization, word cloud, and sentiment analysis approaches are usually used to analyze the data
Data sources include Kaggle, data.gov, Harvard Dataverse, etc.	Data sources include web pages, social media, emails, blogs, journal articles, ETDs, etc.

5. **Knowledge Usage**: "If the knowledge from the post-processing step meets the process objectives, it is put available to the users. Otherwise, another cycle is performed, making changes in the data preparation activities and/or in pattern extraction parameters" [19].

1.3.1   Text Characteristics

Text mining cannot be performed without access to a vast amount of data. This data could be from everyday life or a scientific experiment. In view of this, data management is critical to text mining that requires transforming textual data into a structured form for further analyses using the following:

- NLP for semantic analysis, syntactic parsing, named entity recognition, automatic summarization, parts-of-speech tagging, machine translation, etc.
- Statistical data processing (numerical or language), for instance, to find trends and tendencies
- Advanced pattern recognition algorithms that examine vast collections of data to determine formerly unknown information

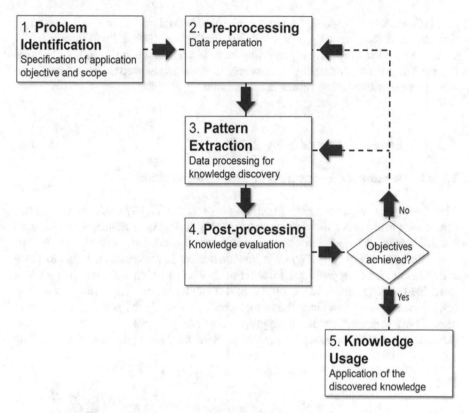

Fig. 1.3 Flow diagram showing text mining process (©2017 Springer Nature, all rights reserved—reprinted with permission from Springer Nature, published in Sinoara et al. [19])

- Data clustering to segregate documents based on their similarities and dissimilarities
- Machine learning algorithms for knowledge discovery

In text mining, "*words* are attributes or predictors and *documents* are cases or records, together these form a sample of data that can feed in well-known learning methods" [20]. The initial presentation of data in a text mining process includes a combination of words, sentences, and paragraphs. It is processed into numerical data leading to the tabular format used in data mining, where a "row represents a document and column represents measurements taken to indicate the presence or absence of words for all the rows without understanding specific property of text such as concepts of grammar or meaning of words" [20].

Given "a collection of documents, a set of attributes will be the total set of *unique words* in the collection called as *dictionary*. For thousands or millions of documents, the dictionary will converge to a smaller number of words" [20]. The technical documents with alphanumeric terms may lead to very large dictionaries, where the tabular layout can become too big in size. "For a large collection of documents, the tabular/matrix layout would be too sparse as individual documents will use only a tiny subset of the potential set of words in a dictionary" [20] and will have lots of 0 values referred to as *sparsing*. A missing value is a non-issue in text mining because the cells are merely indicating the absence or presence of words, and typically we will not deal with missing values in text mining.

1.3.2   Different Text Mining Tasks

1.3.2.1   Document Classification or Text Categorization

The main idea of *document classification* or *text categorization* is to classify documents into one of the given categories (Fig. 1.4). For instance, it can be used for folder/file management where the documents can be organized into folders, one folder for every topic. When a new document is presented, it helps to place a document in the appropriate folder. For this text mining task, we will build a machine learning model based on the past documents and their known categories (elaborated in Chap. 8). Once the model learns to identify the pattern, we can then classify a new document into the appropriate folder. This task can further be broken down into a *binary classification problem* because a given document can belong

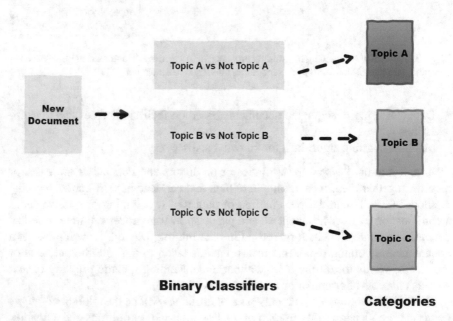

Fig. 1.4 Diagram showing document classification/text categorization

to more than one folder or category (see Sect. 8.1.5). This task is also considered a form of *indexing*, where we assign new documents to their appropriate folder. "For example, automatically forwarding emails to the appropriate department or detecting spam emails" [20]. Typically, a library has different departments that deal with different kinds of work. When a new complaint from a user comes, we can automatically tag the complaint with the department category based on the previously trained model. This task will help improve the efficiency of the library.

1.3.2.2 Information Retrieval

This task deals with matching a particular document with a set of other relevant documents (Fig. 1.5). Suppose you have an extensive collection of documents and you want to find a relevant document. This text mining task retrieves the relevant documents based on the best matches of input document with the collection of documents based on a similarity measure (see Sect. 3.4.9).

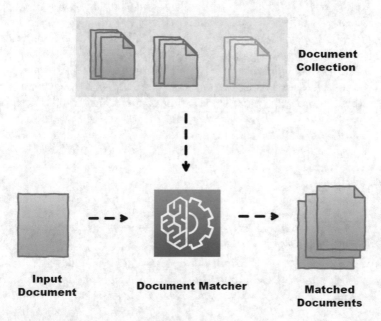

Fig. 1.5 Diagram showing information retrieval

Parthasarathi Mukhopadhyay is a Professor at the Department of Library and Information Science, University of Kalyani.

Story: Designing Enhanced Library OPAC Through Text Analytics Tools—Integration of Koha and Voyant

An OPAC is considered as the mirror of a library collection of any type or size. It has continued to evolve from card catalog to WebOPAC through five different generations. This storyline aims to share an experience in enhancing the capabilities of a WebOPAC through the integration of an open-source ILS and an open-source text analysis tool.

Imagine a library OPAC, where users can search metadata and retrieve real-time item level status (as they do presently in a typical OPAC), can read full text of the retrieved book (if available in an open-access repository) in a book reader (just like they do in Kindle), and can visualize interpretation of texts or corpus of a retrieved document (term links, term circus, term context, document summary, phrases, term context, trends, term correlation, etc. and many such services) in the same OPAC interface with a control to navigate from one text analysis service to another (not presently part of a typical WebOPAC). Moreover, the enhanced features as described can be extended to any Unicode-compliant corpus available in Indic scripts, including tribal languages of India with its own scripts (like Santali in Ol Chiki script).

The rapid progression of "NLP tools and techniques, availability of open-source text analytics tools/services and increasing use of open-source software in library management are the factors that provide tremendous opportunities to enhance a library OPAC" [a] through on-the-fly integration with an appropriate text analytics services. The entire framework for enhanced OPAC requires three components, namely, an ILS with Web-enabled OPAC, a text analytics tool that can produce unique URLs for different text analysis services, and a book reader (optional but it can offer systematic reading of a retrieved document inside an OPAC).

As a proof of concept, this story demonstrates the development of a framework achieved through seamless integration of Koha (as the front-end ILS, koha-community.org), Voyant (as a back-end text analytics tool, voyant-tools.org), and book readers from MagicScroll (magicscroll.net) and Internet Archive (archive.org). The linking mechanism has been crafted carefully so that the "future upgradation of the selected ILS and the selected language" [a] analysis tool would not affect the text analysis services in OPAC. The enhanced framework depends on "URL based linking between Koha catalog dataset, and its corpus in Voyant. The corpus of a Unicode-compliant full-text document at Voyant end (related to a MARC record in Koha) must produce unique URLs for available analysis results. A repeatable MARC tag/subfield needs to be selected (Tag 856 $u is the most logical choice), which can record

(continued)

the corpus data URL (analysis service URL) along with other links. The linking script at Koha's end needs to create a Text Analytics *tab conditionally in OPAC. It means that if MARC tag/subfield provides an analysis URL, then the script will fetch data visualization from the corresponding corpus available at Voyant end in the given OPAC tab with the facility to navigate in other analysis URLs" [a]. The entire mechanisms for developing enhanced OPAC with text analytics features are explained in detail in the slide show at https://textmining-infopros.github.io/collection/chapter1/03-story/.*

Reference:
[a] Mukhopadhyay P, Dutta A (2020) Language analysis in library OPAC designing an open source software based framework for bibliographic records in mainstream and tribal languages. DESIDOC J Library Inf Tech 40:277–285. https://doi.org/10.14429/djlit.40.05.16034

1.3.2.3   Clustering and Organizing Documents

Sometimes we might not be familiar with the labels or categories for a given collection of documents, such that the document structure might be unknown to us. This text mining task helps us cluster similar documents in the collection and assign labels to each cluster (Fig. 1.6). Once the labeling is done, then you know the collection of documents that can be used for document classification. For instance, for a collection of help-desk complaints, librarians can perform clustering and identify the categories and types of complaints.

1.3.2.4   Information Extraction

This task is used to extract data from an unstructured format based on words which can be higher-level concepts or real-valued variables (Fig. 1.7). The measured variable "will not have a fixed position in the text and may not be described in the same way in different documents" [20]. For instance, we can extract the desired values from financial statements heaving details about purchasing books and populate the variables in a database.

1.3.3   Supervised vs. Unsupervised Learning Methods

"Text mining can be divided into supervised and unsupervised learning methods for prediction modeling. Unsupervised learning methods try to find hidden structures out of unlabeled data. They do not need a training phase for the learning of the

Fig. 1.6 Diagram showing
clustering

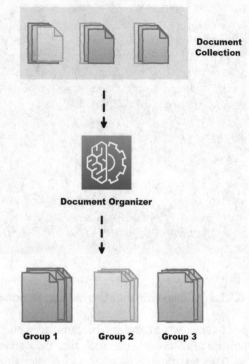

Document
Collection

Document Organizer

Group 1   Group 2   Group 3

Fig. 1.7 Diagram showing
information extraction

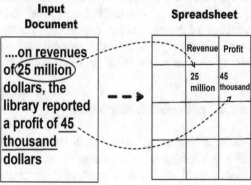

model, therefore can be applied to any text data. Clustering and topic modeling are
commonly used unsupervised learning algorithms in text analysis" [21]. In contrast,
"supervised learning methods are machine learning techniques which learn from
the training data in order to perform predictions on unseen data" [21]. Both learning
methods have been explored and elaborated more in the book.

1.3.4   Cost, Benefits, and Barriers

JISC conducted a study to determine various costs and benefits associated with text mining, which are enumerated below [18]:

1. Costs include access, transaction, entry, staff, and infrastructure costs.
2. Benefits include efficiency, unlocking hidden information and developing new knowledge, exploring new horizons, improved research and evidence base, and improving the research process and quality.
3. Broader economic and societal benefits were also highlighted, such as cost savings and productivity gains, innovative new service development, new business models, and new medical treatments.

They also identified some barriers and risks involved with text mining, namely, "legal uncertainty, orphaned works, and attribution requirements; entry costs; *noise* in results; document formats; information silos and corpora-specific solutions; lack of transparency; lack of support, infrastructure, and technical knowledge; and lack of critical mass" [18].

1.3.5   Limitations

"Text mining corresponds to a more complex problem than data mining. Text is unstructured (or semi-structured when considering its logical structure reflected by XML tags) in nature. In the database field (structured data), each value must respect a specified domain (e.g., a value for the age must be positive and smaller than 120)" [22]. "Text data are not naturally in a format that is suitable for the pattern extraction, which brings additional challenges to an automatic knowledge discovery process" [19]. In other words, the "semantic elements in a text could be implicit, and natural language is faced with different forms of ambiguities together with the problems generated by the synonymy and polysemy attached to many words or expressions" [22]. During social media mining, various problems like "variations, multilingual content, and unusual or incorrect spelling impact the effectiveness of various text-mining tools" [22].

1.4   Case Study: Clustering of Documents Using Two Different Tools

1A: Orange

Problem   You have a mixture of documents such as newspaper articles, research articles, or even catalog metadata on different subjects/topics and want to segregate them roughly based on their similarity and dissimilarity.

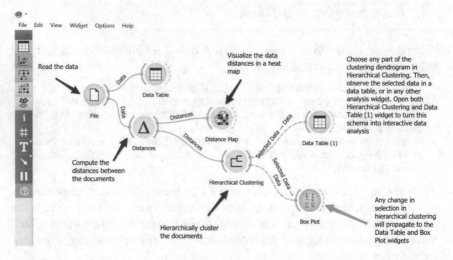

Fig. 1.8 Screenshot showing workflow for hierarchical clustering in Orange

Goal To identify patterns or groups of similar documents within a dataset.

Dataset A CSV file[6] containing two columns with a total of 19 data points. The first column consists of the numbering of documents indicated by S.No., and the other column consists of abstracts from various journal articles.

About the Tool Refer to Chap. 10, Sect. 10.2.5, to know more about `Orange`.

Theory "Clustering is defined as the segmentation of a collection of documents into partitions where documents in the same group (cluster) are more similar to each other than those in other clusters whereas topic modeling is a probabilistic model which is used to determine a soft clustering, where every document has a probability distribution over all the clusters as opposed to hard clustering of documents" [21].

Methodology Figure 1.8 shows the workflow used to perform document clustering in Orange.

The following screenshots demonstrate the steps which were taken to perform clustering:

Step 1: The sample file containing the metadata details of the documents was selected (Fig. 1.9).
Step 2: The file was visualized in a data table (Fig. 1.10).
Step 3: Distances were computed between the rows of the CSV file using normalized Euclidean distance matrix (Fig. 1.11).
Step 4: The computed distances were first visualized in the form of a heatmap (Fig. 1.12).

---

[6] https://github.com/textmining-infopros/chapter1/blob/master/dataset.csv.

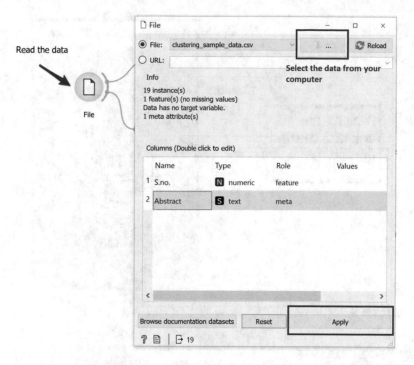

Fig. 1.9 Screenshot showing Step 1

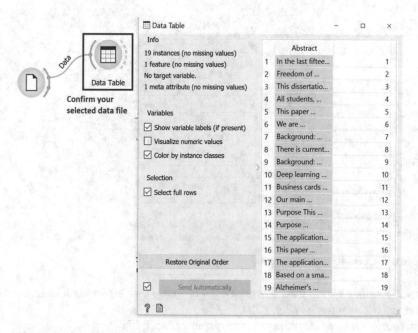

Fig. 1.10 Screenshot showing Step 2

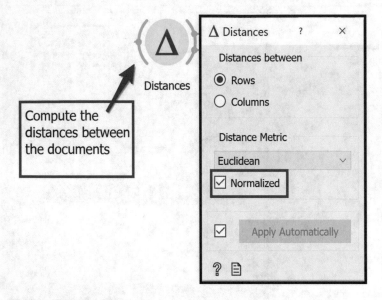

Fig. 1.11 Screenshot showing Step 3

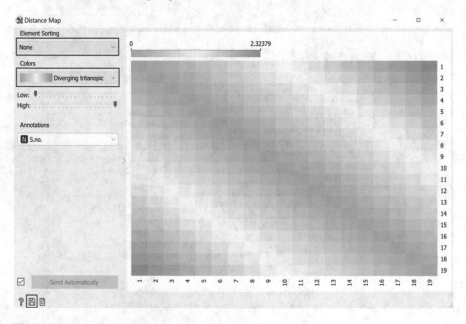

Fig. 1.12 Screenshot showing Step 4

Step 5: Another way to visualize the Step 4 heatmap was to form of a heatmap with clustered ordered leaves (Fig. 1.13).
Step 6: Hierarchical clustering was performed (Fig. 1.14).

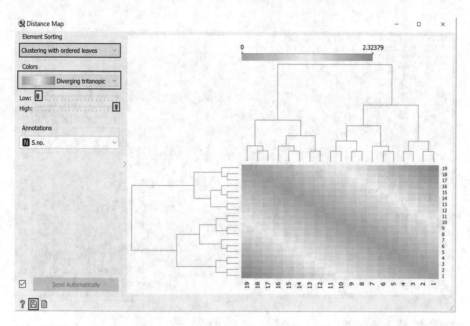

Fig. 1.13 Screenshot showing Step 5

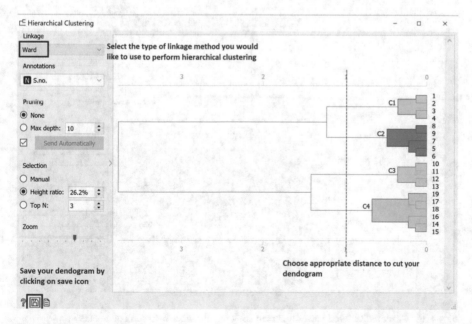

Fig. 1.14 Screenshot showing Step 6

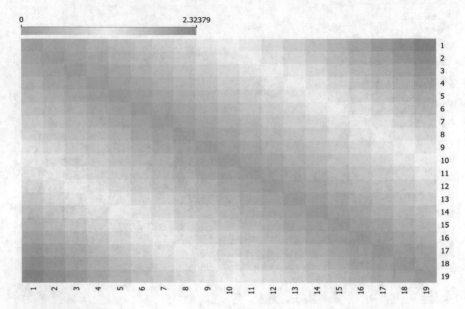

Fig. 1.15 Heatmap showing distances between documents

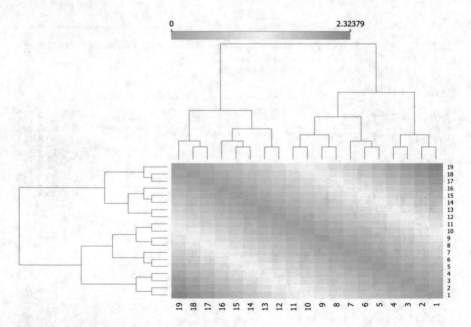

Fig. 1.16 Clustered ordered leaves heatmap showing distances between documents

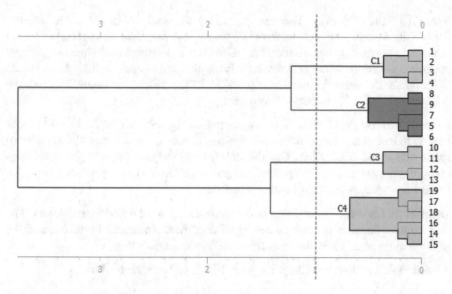

Fig. 1.17 Dendrogram showing hierarchical clustering of documents

Results Figures 1.15 and 1.16 show the heatmap and clustered heatmap plots for distances between the documents using the normalized Euclidean distance metric. Blue-pink gradient color was used to show the correlation between the documents, where pink color showed a high correlation between the documents and blue color showed a low correlation.

Figure 1.17 presents the dendrogram using the ward method. The number of clusters was determined by cutting the dendrogram using a vertical dotted line at the value 1.0, resulting in 4 main clusters (C1, C2, C3, and C4) and 11 sub-clusters. C1 was related to C2, and C3 was related to C4. C1 consisted of documents 1 and 2 as one sub-cluster and documents 3 and 4 as another. Similarly, C2 consisted of documents 8 and 9 as the first sub-cluster, documents 5 and 6 as the second sub-cluster, and document 7 as the third sub-cluster. C3 consisted of documents 10 and 11 as one sub-cluster and documents 12 and 13 as another. In contrast, C4 consisted of document 19 as the first sub-cluster, documents 17 and 18 as the second sub-cluster, document 16 as the third sub-cluster, and documents 14 and 15 as the fourth sub-cluster.

1B: R

Problem Given a mixture of documents such as newspaper articles, research articles, or even catalog metadata on different subject/topics, and you want to segregate them roughly based on their similarity and dissimilarity.

Prerequisite Familiarity with the R language.

Virtual RStudio Server You can reproduce the analysis in the cloud without having to install any software or downloading the data. The computational environment runs using BinderHub. Use the link (https://mybinder.org/v2/gh/textmining-infopros/chapter1/master?urlpath=rstudio) to open an interactive virtual RStudio environment for hands-on practice. In the virtual environment, open the `clustering.R` file to perform clustering.

Virtual Jupyter Notebook You can reproduce the analysis in the cloud without having to install any software or downloading the data. The computational environment runs using BinderHub. Use the link (https://mybinder.org/v2/gh/textmining-infopros/chapter1/master?filepath=Case_Study_1B.ipynb) to open an interactive virtual Jupyter notebook for hands-on practice.

Dataset A CSV file[7] containing two columns with a total of 19 data points. The first column consists of the numbering of documents indicated by S.No., and the other column consists of abstracts from various journal articles.

About the Tool Refer to Chap. 10, Sect. 10.2.1, to know more about R.

Theory "Clustering is defined as the segmentation of a collection of documents into partitions where documents in the same group (cluster) are more similar to each other than those in other clusters whereas topic modeling is a probabilistic model which is used to determine a soft clustering, where every document has a probability distribution over all the clusters as opposed to hard clustering of documents" [21].

Methodology and Results The libraries and the dataset required to perform document clustering in R were loaded.

```
#Load libraries
library(tm)
library(proxy)
library(RTextTools)
library(fpc)
library(wordcloud)
library(cluster)
library(stringi)
library(textmineR)
library(factoextra)
library(readr)
library(ggplot2)
library(igraph)

#Load dataset
data <- read.csv("https://raw.githubusercontent.com/
textmining-infopros/chapter1/master/dataset.csv")
```

---

[7] https://github.com/textmining-infopros/chapter1/blob/master/dataset.csv.

A document-term matrix (DTM) was created, followed by a term frequency-inverse document frequency (TF-IDF) matrix. TF-IDF is determined by multiplying the term frequency (counts of words in documents in our case) by an inverse document frequency (IDF) vector (explained thoroughly in Chap. 3, Sect. 3.4.5). The textmineR package in R calculates IDF for the ith word as:

$$\text{IDF}_i = \ln\left(\frac{N}{\sum_{j=1}^{N} C\left(\text{word}_i, \text{doc}_j\right)}\right),\tag{1.1}$$

"where N is the number of documents in the corpus. By default, when one multiplies a matrix with a vector in R, the vector gets multiplied to each column, due to which DTM is transposed before multiplying the IDF vector and is then back transposed to its original orientation" [23].

```
#Create Document Term Matrix
    dtm <- CreateDtm(doc_vec = data$Abstract,
        doc_names = data$S.No.,
        ngram_window = c(1,2),
        stopword_vec = c(stopwords::stopwords("en"),
                    stopwords::stopwords(source = 'smart')),
        lower = TRUE,
        remove_punctuation = TRUE,
        remove_numbers = TRUE,
        verbose = FALSE,
        cpus = 2)

#Construct matrix of term counts to get IDF vector
    tf_mat <- TermDocFreq(dtm)

#TF-IDF
    tfidf <- t(dtm[ , tf_mat$term ]) * tf_mat$idf
    tfidf <- t(tfidf)

#Convert TF-IDF matrix to standard R matrix
    m <- as.matrix(tfidf)
```

The number of clusters was determined using the *elbow method*, where the bend or knee in the plot is considered as an indicator of an optimal number of clusters for the data. The bend further indicates that any additional number of clusters beyond the bend will have little or no value. It can be observed from Fig. 1.18 that the bend was formed at number 5. Therefore, five clusters were used to fit the data. Further, Fig. 1.18 showed that as the number of clusters (k) increases, the variance decreases.

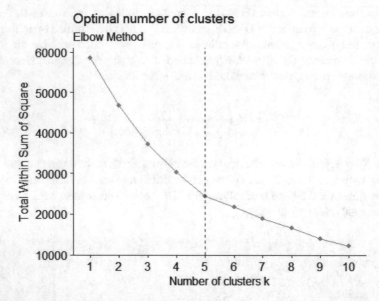

Fig. 1.18 Optimal number of clusters by elbow method

```
#Determine the number of K for clustering
fviz_nbclust(m, kmeans, method = "wss")
fviz_nbclust(m, kmeans, method = "wss") +
  geom_vline(xintercept = 5, linetype = 2) +
  labs(subtitle = "Elbow Method")
```

Euclidean distance method was used to determine the distance between the documents. The hierarchical clustering algorithm uses this distance to cluster the documents.

```
#Compute Distance Matrix
require(stats)
res.dist <- dist(x=m, method = "euclidean")
```

Figure 1.19 represents the distance matrix of the dataset. The red color indicated high similarity, while the blue color indicated low similarity. The color level is proportional to the value of the similarity between documents, where pure red represents 0 and pure blue represents 1. As shown in the figure, documents 2, 19, 13, and 8 were very dissimilar to all the other documents and even to each other (colored in blue) and thus can be represented as different clusters. In contrast, the rest of the

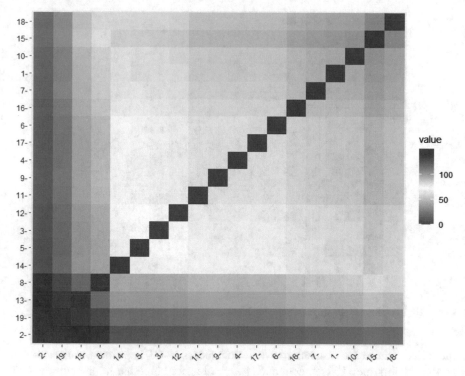

Fig. 1.19 Heatmap showing the distance matrix for documents

documents show little or more similarity to each other and can be clustered together in one group.

#Visualizing distance matrices
```
friz_dist(res.dist)
```

Another way to visualize the distances between the documents is hierarchical clustering with dendrogram. A dendrogram is a tree diagram that shows the hierarchical relationship between the data points (in our case, different documents). Hierarchical clustering was performed using the ward's method, and the tree was cut with five clusters at a distance of 70 (Fig. 1.20). The x-axis of Fig. 1.20 indicates the document's S.No, whereas the y-axis of the figure shows the distance or dissimilarity between the clusters calculated by the Euclidean method. The distance is inversely proportional to the similarity of the documents, which means the smaller the distance between the two documents, the more similar they are. The documents that are close to each other have slight dissimilarity and were linked together. It can be observed from Fig. 1.20, five clusters, viz., C1, C2, C3, C4, and C5 were

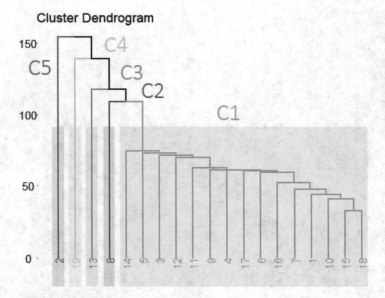

Fig. 1.20 Rectangular dendrogram showing hierarchical clustering of documents

identified, where the distance between documents in C1 kept increasing with the level of the merger. Fifteen documents were clustered together in one cluster (C1) based on their similarity. In contrast, each of the remaining four clusters consisted of just one document, which meant there were more documents on a subject/topic than the others in the dataset.

```
#Agglomerative hierarchical clustering
require(stats)
res.hc <- hclust(d = res.dist, method = "ward.D")
plot(x = res.hc)

#Coloring dendrogram
require(factorextra)
fviz_dend(x=res.hc, cex = 0.7, lwd = 0.7)
require("ggsci")
fviz_dend(x=res.hc, cex = 0.8, lwd = 0.8, k=5, rect = TRUE,
    k_colors = "jco", rect_border = "jco", rect_fill = TRUE,
    ggtheme = theme_void())
```

The dendrogram in hierarchical clustering can be visualized with a rectangular (Fig. 1.20), circular (Fig. 1.21a), or phylogenic (Fig. 1.21b) structure. It is a way of visualizing the same results with a different perspective according to your research problem and dataset.

(a)

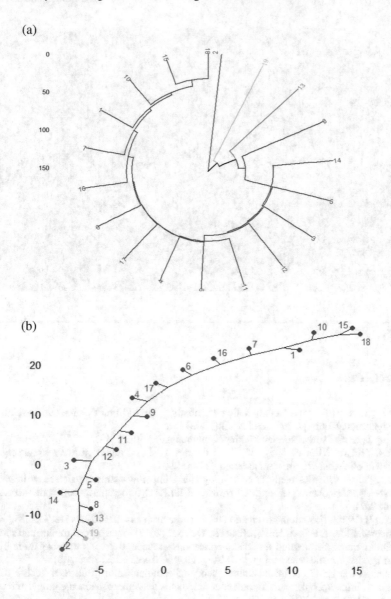

(b)

Fig. 1.21 Different types of dendrograms. (**a**) Circular dendrogram showing hierarchical clustering of documents. (**b**) Phylogenic dendrogram showing hierarchical clustering of documents

#Different types of dendrograms
```
fviz_dend(res.hc, cex = 0.8, lwd = 0.8, k= 4,
   rect = TRUE,
   k_colors = "jco",
   rect_border = "jco",
   rect_fill = TRUE,
   type = "circular",
   repel = TRUE)

fviz_dend(res.hc, cex = 0.8, lwd = 0.8, k= 4,
   rect = TRUE,
   k_colors = "jco",
   rect_border = "jco",
   rect_fill = TRUE,
   type = "phylogenic",
   repel = TRUE)
```

References

1. Fay   B   (2019)   The   computational   library.   https://libraries.mit.edu/news/bibliotech/computational-library/. Accessed 21 June 2021
2. Wing J (2006) Computational thinking. Commun ACM 49(3):33–35
3. ALA (2018) What & why of CT. Libraries ready to code https://www.ala.org/tools/readytocode/what-why-of-ct. Accessed 22 June 2021
4. ALA (2018) Libraries ready to code Infographic. https://www.ala.org/advocacy/sites/ala.org.advocacy/files/content/pdfs/draft%20v6%20-%20LRTC_infographic%20(2).pdf. Accessed 25 June 2021
5. Luhn H (1958) The automatic creation of literature abstracts. IBM J Res Dev 2(2):159–165
6. Miner G, Elder J, Fast A, Hill T, Nisbet R, Delen D (2012) Practical text mining and statistical analysis for non-structured text data applications. Academic Press. https://www.oreilly.com/library/view/practical-text-mining/9780123869791/. Accessed 22 June 2021
7. Shannon CE (1948) A mathematical theory of communication. Bell Syst Tech J 27:379–423.   https://people.math.harvard.edu/~ctm/home/text/others/shannon/entropy/entropy.pdf. Accessed 22 June 2021
8. Bates M (1995) Models of natural language understanding. PNAS 92:9977. https://doi.org/10.1073/pnas.92.22.9977
9. Cutting DR, Karger DR, Pederson JO, Tukey JW (1992) Scatter/Gather: a cluster-based approach in browsing large document collections. In: Proc 15th annual international ACM-SIGIR conference of research and development in information retrieval, Copenhagen. ACM Press, New York, pp 318–329
10. Tombros Villa R, Rijsbergen CJ (2002) The effectiveness of query-specific hierarchic clustering in information retrieval. Inf Process Manag 38(4):559–582

11. Salton G, Allan J, Buckley C, Singhal A (1996) Automatic Analysis, Theme Generation, and Summarization of Machine-Readable Texts. In: Agosti M, Smeaton AF (eds) Information retrieval and hypertext. Springer, Boston, MA. https://doi.org/10.1007/978-1-4613-1373-1_3

12. Anoopa VS, Asharafa S, Deepak P (2016) Unsupervised concept hierarchy learning: A topic modeling guided approach. Procedia Comput Sci 89:386–394. https://doi.org/10.1016/j.procs.2016.06.086

13. Ranganathan SR (1931) The five laws of library science. Madras Library Association, Madras

14. Nicholson S (2003) The bibliomining process: Data warehousing and data mining for library decision making. Inf Technol Libr 22(4):146–150

15. Lammey R (2015) CrossRef text and data mining services. Insights 28:62–68. https://doi.org/10.1629/uksg.233

16. Lamba M, Madhusdhuan M (2018) Application of topic mining and prediction modeling tools for library and information science journals. In: Library practices in digital era: Festschrift in Honor of Prof. V Vishwa Mohan. BS Publications, Hyderabad, pp 395–401. https://doi.org/10.5281/zenodo.1298739

17. Salloum SA, Al-Emran M, Monem AA, Shaalan K (2018) Using text mining techniques for extracting information from research articles. In: Shaalan K, Hassanien AE, Tolba F (eds) Intelligent natural language processing: Trends and applications. Springer International Publishing, Cham, pp 373–397

18. Reimer T (2012) Key issue: Text mining, copyright and the benefits and barriers to innovation. Insights 25:212–215. https://doi.org/10.1629/2048-7754.25.2.212

19. Sinoara RA, Antunes J, Rezende SO (2017) Text mining and semantics: a systematic mapping study. J Braz Comput Soc 23:9. https://doi.org/10.1186/s13173-017-0058-7

20. Weiss SM, Indurkhya N, Zhang T, Damerau FJ (2005) Overview of text mining. In: Weiss SM, Indurkhya N, Zhang T, Damerau FJ (eds) Text mining: Predictive methods for analyzing unstructured information. Springer, New York, NY, pp 1–13

21. Allahyari M, Pouriyeh S, Assefi M, et al (2017) A brief survey of text mining: Classification, clustering and extraction techniques. https://arxiv.org/pdf/1707.02919.pdf. Accessed 23 June 2021

22. Savoy J (2018) Working with text. Tools, techniques and approaches for text mining. In: Tonkin EL, Tourte GJL (eds). Chandos Publisher, Cambridge, MA, 2016, 330 pp. (ISBN 978-1-84334-749-1). J Assoc Inf Sci Tech 69:181–184. https://doi.org/10.1002/asi.23899

23. Jones TW (2019) Document clustering. https://cran.ms.unimelb.edu.au/web/packages/textmineR/vignettes/b_document_clustering.html. Accessed 22 June 2021

Additional Resources

1. Padilla T, Allen L, Frost H, et al (2019) Always already computational: Collections as data—Final report. https://zenodo.org/record/3152935#.X6WOf-LPzIU. Accessed 25 June 2021

Chapter 2
Text Data and Where to Find Them?

Abstract This chapter first throws light on the standard data file types with their usage, advantages, and disadvantages. In a digital library, data might be useless and considered incomplete without a metadata record. Therefore, the functions, uses, components, and importance of metadata are covered comprehensively, followed by steps to create quality metadata, common metadata standards available, different metadata repositories, common concerns, and solutions. The second part of the chapter focuses on the importance of the inclusion of optical character recognition (OCR) for digitized data, followed by different ways of getting data from (i) online repositories, (ii) relational databases, (iii) web APIs, and (iv) web/screen scraping to start a text mining project. Further, several online repositories, language corpora, and repositories with APIs available for text mining are enumerated. Finally, some of the essential applications of APIs for librarians and for what purpose librarians can use them in their day-to-day work are covered in this chapter.

2.1 Data

Data is a product of research that is heterogeneous across and contextualized within the academic discipline. It is the fundamental unit of information. There is no universal definition of data, and it differs from discipline to discipline. Still, one can define it as the records essential for reproducing and assessing research results, as well as the events and processes leading to those results. It can be quantitative or qualitative values or pieces of information that should be (i) valid, (ii) shared, (ii) heterogeneous, and (iv) controlled in a research community. Data is more than just facts, as they are authored objects created for an audience. Datasets have some characters as text. They are susceptible to heuristic analysis, where the results are heuristics and are a construct that is created.

Digital data is electronic information that a computer can understand, for instance, electronic images, texts, and structured data like spreadsheets. A few examples of data include DNA samples, specimens, geospatial coordinates, databases, codes/algorithms, models, citations, digital tools, documentation, CSV files, 3D models, JPEG files, and machine-readable files.

Data can be textual or non-textual. Textual data are originated from various digital sources such as blogs, social networks, or forums. Textual data is a form of unstructured data that humans can easily interpret and process, but it is difficult for machines to perform the same. Products like sites, blogs, posters, reports, conferences, questionnaires, articles, codebooks, white papers, and books help make the data meaningful. These products are called background data as they contain the contextual information in analyzing the primary foreground data collected for the study. The analyses of textual data can help us to provide answers to some of the essential questions:

1. What words/phrases/topics are statically significant? What are outliers?
2. How are these dispersed through the corpus?
3. How are words or topics related to sentiment or style in the text?
4. What words or topics relate more closely with others?
5. How similar is one text to another text?
6. What proper names (such as people, places) appear in the text and where?
7. How does the focus of a corpus change over time or geography?

2.1.1   Digital Trace Data

Digital trace data (also referred to as digital footprint or digital breadcrumbs) is the unprecedented amount of data generated from blogs, web search data, social media sites, other Internet forums, administrative data on sites, Internet archive, digitized text data archive, audiovisual data, telecommunication, or geospatial data. This voluminous amount of digital data generated on the Internet helps map human behavior and how we interact in different groups of communities and societies. It also opens the gates to perform various analytics that can then be used to do some social good. In 2010, Eagle et al. [1] were the first to conduct a study that used what we call today digital trace data or digital footprint.

"With scholarship moving online, we can now access various types of altmetrics digital traces such as reading, organizing, sharing, and discussing scientific papers" [2]. Altmetrics is the data that captures engagement with research that happens across the web. As people share research on social media, in the news, or in public policy documents, all of this engagement with research online leaves digital traces on the web. In an altmetrics analysis, two types of data can be processed: (i) *metrics* and (ii) *qualitative*. *Metrics* data help to analyze the research engagement and quantify how often research is being shared, how many tweets/posts have been shared, and how it correlates with citations. In contrast, *qualitative* data helps to understand who is talking about a particular research, where they are talking about it, what they are saying, and other information that contextualizes the engagement. Further, in order to determine the meaning of *qualitative* data, you need to zoom out from reading individual mentions to look out the common themes (see Chaps. 4 and 6) or sentiments (see Chap. 7) across many different altmetrics posts.

2.1.1.1 Features or Strengths of Digital Trace Data

The major features of digital trace data are as follows [3]:

1. *Volume*: It is around us everywhere and is increasing rapidly.
2. *Always On*: It is continuously being collected, unlike data collected by surveys which might miss significant events taking place around the world, such as pandemics, protests, revolutions, or market surges, if they are not executed at the right time and right place.
3. *Nonreactive*: It is unreactive and non-interactive in comparison to other techniques such as interviews or surveys due to the social desirability bias or other forms of biasness by interviewee.
4. *Captures Social Relationships*: It measures social relationships such as a network of followers and following on Twitter or public reviewing new books on Goodreads.

2.1.1.2 Weaknesses of Digital Trace Data

The major weaknesses of digital trace data are as follows [3]:

1. *Incompleteness*: Irrespective of the size and scale of digital trace data, the amount of data that is either missing or incomplete is still quite huge.
2. *Inaccessibility*: Most of the government and industrial data are inaccessible to researchers. Further, where Twitter is a public platform whose data is publicly available for analyses, the vast majority of platforms, such as Instagram, Facebook, and Quora, have restricted access to their data.
3. *Non-Representativeness*: Many times when working with a random sample of data generated by a public platform, such as Twitter or Facebook, the data is not representative of their broader population, but within-sample comparisons, non-representative data can be potent, as long as researchers are clear about the characteristics of their sample and support with theoretical or empirical evidence.
4. *Driftness*: With the growth in the number of applications and platforms generated every year, there is a risk of driftness of users from one platform to another with time. In other words, the digital platform which is popular today may become obsolete in the coming months/years as a new better interactive platform becomes popular among the users and can cause a driftness effect, where the users leave the previous platform or become more active for the newer one.
5. *Algorithmic Confounding*: The interference of algorithms with the data to cause a perception of false/biased results. Elaborated more in Chap. 11, Sect. 11.3.3.
6. *Unstructured*: It is very disordered, difficult to search, difficult to convert, and difficult to transpose across different formats.
7. *Sensitivity*: It often contains sensitive information and needs to be handled with care during text pre-processing and cleaning steps so that the identity and demographics are not revealed for the users.

8. *Positivity-Biasness*: On digital platforms, users usually share positive news or give positive comments due to the presence of their friends, colleagues, or other people in their friends/followers' lists and often avoid reporting negative information/comments. It usually creates a common form of biasness in social media research.

Digital trace data is indispensable to librarians. It is found all around them in the form of digitized print resources, journals, electronic/print theses and dissertations, newspapers, virtual librarian chat archives, comments/reviews on WebOPACs, library social media accounts, administrative records, reference desk queries, and so on. In this book, text/textual data will be reversibly used as digital trace data.

2.1.1.3 Use Case: Bibliomining

In 2003, Nicholson and Stanton [4] first used a term called *bibliomining* to discuss the approach of data mining in libraries where the data mining and bibliometric tools are applied to the data produced from library services. "The term pays homage to bibliometrics, which is the science of pattern discovery in scientific communication. It uncovers the data of usage patterns in order to provide tailor-made services to meet the needs of different library's user groups" (Nicholson [4]). Bibliomining analyzes operations such as circulation, purchases, and cataloging and finds trends in recorded information from user profiles. Libraries manage large stores of data about collections and usage in their library management systems (LMS). It helps to identify the pattern from a vast collection of data associated with LMS to help in the decision-making process or justify the services provided by the library. Basic steps in a bibliomining process (Fig. 2.1) consist of the following [4]:

1. *Determination of areas of focus*: Defining the specific problem in the library or maybe a general problem that requires exploration and decision-making.
2. *Identification of internal and external data sources*: Wherein library management system (LMS) is the primary source for bibliomining. The two additional sources that can be used to get the data for bibliomining analysis are (i) internal data sources of the library, which are already within the LMS, such as transactional data, patron database, and web server logs, and (ii) external data sources, which are not located within the library system, such as demographic information related to a specific ID number located in the personnel management system.
3. *Creating a data warehouse*: The cleaned and anonymized version of the operational data is reformatted for bibliomining analysis. It corresponds to the collection, cleaning, and anonymizing or extraction, transformation, and loading (ETL) of data into the data warehouse.
4. *Selection of tools for analysis.*
5. *Discovery of patterns.*
6. *Visualizing or presenting the discovered trend or pattern.*
7. *Implementing the results to provide tailor-made services to the users or make an appropriate decision based on the results.*

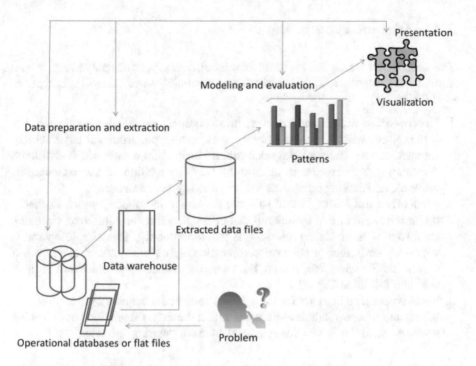

Fig. 2.1 Process of bibliomining

There are numerous benefits to libraries and librarians of using bibliomining in libraries. These include the following:

1. Providing efficient library services like current awareness services (CAS) or selective dissemination of information (SDI) to its patrons.
2. Weeding out of books which are having low usage pattern over the years.
3. Deciding collection development, planning, budgeting, and purchasing of resources.
4. Identifying the demographics of active clients, the relationship between the clients and the materials demanded, and the readership pattern of the clients.
5. Weblog files provide a pattern for what, how, and when the library sites are being used. Data mining on weblogs will help to improve the library sites.
6. Acting as a support tool in various decision-making processes.
7. Determining the rush hours of the library.
8. Recognizing the number of downloads from research articles, thus, can help in determining core journals and can help in weeding out the journals with a minimum number of downloads/usage.

2.2   Different Types of Data

The structure of words and numbers enables different types of analysis either by a human or a computer reading the text. There are mainly three categories or types of data (Table 2.1):

1. **Unstructured data** is not organized into a distinct predefined semantic unit. For instance, any written account, literary work, newspaper article, or anything represented as text. Human beings can process unstructured data, which is difficult for computers to process in meaningful ways by reading it, understanding its contents, and making inferences using context and common sense.
2. **Structured data** is organized to some particular data model, which explicitly defines the structure of the data. It can be processed by any software, assuming each row of the table represents one individual entity. This simple operation would be far harder to perform computationally on unstructured data, even if it contains the same information. For example, data about presidents of India in a tabular format in Table 2.2.
3. **Semi-structured** does not follow a formal data model but uses formal constructs to indicate separate semantic elements within the data. One method of indicating semantic elements is by adding special codes or "markup" to what is otherwise

Table 2.1 Summary of different types of data

Category	Definition	Advantages	Disadvantages	Examples
Unstructured data	Data that is not organized into distinct predefined units	1. Usually human-readable 2. Semantically rich	Difficult for software to process, analyze, and interpret	Textual data, including essays, newspaper articles, and literary works
Structured data	Data that is organized according to an explicit, well-defined model	Easy for software to process and analyze	Structure may limit types of data that can be represented	1. Tabular data like spreadsheets 2. Image files like TIFF or JPEG 3. Binary files
semi-structured data	Data that uses formal constructs, such as markup codes to indicate separate semantic elements	1. Can support software processing of unstructured data 2. Can preserve the semantic rich-ness of the unstructured text	Moderately challenging to process using software	1. HTML documents 2. XML documents

Table 2.2 Example of structured data

Name	Birth year	Elected	Left office
Ram Nath Kovid	1945	2017	2022
Pranab Mukherjee	1935	2012	2017
Pratibha Patil	1934	2007	2012
A.P.J. Abdul Kalam	1931	2002	2007

unstructured text. This approach avoids ambiguities and appeals to context and common sense when processing content in natural languages like English. For example, adding codes "<name>" and "</name>" around the occurrence of a proper name.

2.3 Data File Types

Data can be stored and expressed in a wide range of formats. Each format is best suited for unstructured, semi-structured, or structured data. File types can be separated into two classes: text-based formats and binary (i.e., non-text-based) formats. A text file is machine-readable through character encoding standards such as ASCII and Unicode, whereas a binary file runs through applicative software and may be proprietary. Some file types include both binary and text, for example, rich text format. A file format is a method of encoding information within a computer file. A program or an application might need to recognize the file format to access the content within the file. Therefore, it is crucial to know the type of format one will choose for the text mining process. A file format is often indicated as part of the file name in an extension or suffix. The extension usually follows a period in the file name, for instance, .jpg, .txt, .docx, etc. Files in proprietary formats have to be opened by specific software in which they have been created. Therefore, for long-term preservation of data, one should either use open and nonproprietary file formats or convert their data to open standards.

2.3.1 Plain Text

In computational processing, each letter and punctuation of a natural language like English is called a character. A key feature of character is that it is abstract to its style, color, and font, typically when we process text on a computer. For instance, when we look for "dog pictures" in an Internet search, we do not care whether the word "dog" happened to be written in bold or italics or a particular font or color. Since computers internally operate on data represented by numbers, the simplest way to represent characters is by assigning a unique number to each distinct abstract form. The standard scheme for assigning these numbers to abstract forms is called Unicode. Unicode aims to standardize and assign different numbers to all characters

Character	Unicode number
a a *a* a ɑ	97
b **b** *b* b b	98
z **z** *z* z z	122
ȸ ȸ ȸ ȸ ȸ	0238
Ȅ Ȅ Ȅ Ȅ Ȅ	0204

Fig. 2.2 Showing Unicode for different character

of all commonly used writing systems. Until now, over 80,000 numbers have been assigned using Unicode. Consider Fig. 2.2. With individual characters represented in this way, it is easy to represent the text of any length by merely arranging their corresponding numbers together in sequence, forming one character after the other. In addition to spaces, line break (which indicates that the text occurs after it should appear at the start of the line below the last character) and punctuation marks (such as periods or commas) are also considered characters. They are given unique numbers so that they can be used to designate additional relationships between alphabetic characters and words.

There is one complication as a result of the long history of computer processing of text. Early computer systems represented characters using small units of memory called bytes. Each byte can represent one of 256 possible states. Since many of the earliest computer systems processed only English text, these 256 states were sufficient to directly represent all of the commonly used letters and symbols needed at the time. Hence, conventions and standards emerged based upon using one single byte to represent a character. Modern software, by contrast, is often expected to be capable of processing characters in any language—a task that requires more than the allocated 256 possible states in a byte. Encoding is used to express Unicode characters as sequences of multiple bytes. The most commonly used encoding for modern computer systems is the *UTF-8*, as it can represent any Unicode character.

2.3.1.1 Typical Uses

Plain text files are widely used for storing and exchanging unformatted text between programs. They are also commonly used in many text mining applications, in which plain text is often a starting point for textual analysis.

2.3.1.2 Advantages

1. It contains encoded content representing a sequence of characters.
2. It is one of the most used and flexible file types.

3. Many other file formats are built upon the plain text by using plain text format to represent their contents.
4. They are widely supported by virtually all software capable of processing text, such as word processors, web browsers, and countless utility programs.
5. Creating or saving research data in text format helps one read that file in a plain text editor and is also human-readable.

2.3.1.3   Disadvantages

1. It does not provide any further information about how the content is structured or how particular parts of it should be interpreted or processed; that is, a program cannot further process a plain text file's contents except as a sequence of characters.
2. Other file formats also impose new conventions on how the textual contents can be structured.

2.3.2   CSV

Comma-separated value (CSV) is a file format built upon the foundation of plain text. Therefore, it can be opened in a text editor, word processor, or other application capable of processing plain text. CSV is generally used to represent tabular data, i.e., the data is divided into rows and columns. Data in a CSV file, therefore, has a fixed structure imposed upon it. While it may contain any amount of text, all of this text must be filed into separate boxes or "cells" arranged in a single two-dimensional grid. CSV represents this by imposing simple conventions on a plain text file (Figs. 2.3 and 2.4):

1. A comma is used to separate a cell from the next cell to the right.
2. A line break is used to separate one row from the next row below.
3. Other symbols (usually quotation marks) may be used to avoid ambiguity where the text in a cell contains a line break, comma, or other characters with special meaning.

2.3.2.1   Typical Uses

In many everyday use cases, it represents a single logical table of data. The first row contains column headings, and each successive row contains a series of data about one item, each filed under the relevant heading.

	A	B	C	D	E	F	G	H	I	J
row no	id	created a	from user	from user id	to user	to user id	language	source	text
1	1.22485E+18	Wed Feb	The Econc	5988062	?		-1	en	SocialFloy	Globally, road accidents kill more people every year
2	1.22451E+18	Tue Feb 0	QuickTake	252751061	?		-1	en	SocialFloy	Drones on the island of Zanzibar are spraying a silic
3	1.22475E+18	Tue Feb 0	Dr. Joachi	41779021	?		-1	en	Twitter fo	Did you know that #Cancer was responsible for mo
4	1.22493E+18	Wed Feb	Ashiembi	246075279	?		-1	en	Twitter fo	RT @Asamoh_: Moi was MP for 45 years. He was p
5	1.22493E+18	Wed Feb	Eightball8	1.1869E+18	realDonal	25073877		en	Twitter fo	@realDonaldTrump Nice game show! Malaria, shov
6	1.22493E+18	Wed Feb	Mummy's	1.01543E+18	?		-1	en	Twitter fo	RT @AsiwajuLerry: Girls having mixed feelings, sadr
7	1.22493E+18	Wed Feb	kristofa Tc	576550026	?		-1	en	Twitter fo	RT @juditrius: Learning at the south-south meeting
8	1.22493E+18	Wed Feb	nagapuri	1440517916	?		-1	en	Twitter fo	We are facing health issues due to buffaloes on Stre
9	1.22493E+18	Wed Feb	ANYONE I	1.03375E+18	Jonathan	94784682		en	Twitter fo	@JonathanTurley Yet you're okay with the Cheeto k
10	1.22493E+18	Wed Feb	Dude. She	4248168732	Sahilcdes:	7.82312E+17		en	Twitter fo	@SahilcdesaiC Rush von Opioidfiend is clearly close
11	1.22493E+18	Wed Feb	ArsenalTr	9.13112E+17	?		-1	en	Twitter fo	RT @Asamoh_: Moi was MP for 45 years. He was p
12	1.22493E+18	Wed Feb	babe.	1.09757E+18	?		-1	en	Twitter fo	RT @AsiwajuLerry: Girls having mixed feelings, sadr
13	1.22492E+18	Wed Feb	Sandy H	3278686212	ActuallyJ	585117688		en	Twitter fo	@ActuallyJ @gngavon @DrGJackBrown Malaria - tl
14	1.22492E+18	Wed Feb	DP 👊🐍	8.82393E+17	realTuckF	55060090		en	Twitter fo	@realTuckFrumper Oh and nude Malaria pictures a
15	1.22492E+18	Wed Feb	ActuallyJ	585117688	gngavon	7.88962E+17		en	Twitter W	@gngavon @SandyH1123 @DrGJackBrown He didr
16	1.22492E+18	Wed Feb	Good Wit	350371379	?		-1	en	Twitter fo	RT @naughtywriter2: I felt like we were watching S
17	1.22492E+18	Wed Feb	Wavyson	1.03181E+18	?		-1	en	Twitter fo	RT @benxta_: omg this reminds me of the time i to
18	1.22492E+18	Wed Feb	Honeysch	358357700	EdanClay	7.93903E+17		en	Twitter fo	@EdanClay @SpeakerPelosi Malaria looks like: Oh,
19	1.22492E+18	Wed Feb	Leah	515912573	?		-1	en	Twitter W	RT @naughtywriter2: I felt like we were watching S
20	1.22492E+18	Wed Feb	Dude. She	4248168732	charliekir	292929271		en	Twitter fo	@charliekirk11 @DonaldJTrumpJr She was preppin
21	1.22492E+18	Wed Feb	Mazi Pete	1571795894	cchukude	266867464		en	Twitter fo	@cchukudebelu When pple eat live frog 🐸 and all
22	1.22492E+18	Wed Feb	Myandra	595582707	?		-1	en	Twitter W	I felt like we were watching Satan's version of The E
23	1.22492E+18	Wed Feb	Dr.K'OLOC	2767438544	?		-1	en	Twitter fo	RT @Asamoh_: Moi was MP for 45 years. He was p
24	1.22492E+18	Wed Feb	danyela 👒	8.97654E+17	?		-1	en	Twitter fo	RT @AsiwajuLerry: Girls having mixed feelings, sadr
25	1.22492E+18	Wed Feb	Somya Mi	9.13303E+17	?		-1	en	Twitter fo	RT @MelanieEmmajade: Malaria is complicated an
26	1.22492E+18	Wed Feb	Favour	8.189E+17	?		-1	en	Twitter fo	RT @AsiwajuLerry: Girls having mixed feelings, sadr
27	1.22492E+18	Wed Feb	Sumit Guj	60954154	?		-1	en	Twitter fo	RT @TheSanjivKapoor: Some perspective: In the pa
28	1.22492E+18	Wed Feb	Somya Mi	9.13303E+17	?		-1	en	Twitter fo	RT @rihickson: .@somya_mehra from @unimelb @
29	1.22492E+18	Wed Feb	Kenyan Tr	8.75763E+17	?		-1	en	Kenyan Tr	LETTERS: Why recurrent malaria outbreaks are wor

Fig. 2.3   Example of tweets for malaria query in a spreadsheet application

2.3.2.2   Advantages

CSV is a straightforward format to create and process, and it has an unambiguous structure.

2.3.2.3   Disadvantages

1. The rigid structure of CSV, where only one single table of data is represented. Any further structure, for example, additional information within each cell, cannot be represented.
2. Many programs are incompatible with their format.
3. It does not provide a mechanism for recording which text encoding is used in the file, leading to data corruption when an incorrect encoding is used to decode file contents.

2.3.3   JSON

JavaScript Object Notation (JSON) is a file format built upon the foundation of plain text. JSON files represent information logically composed in two types of structure: (a) *lists of items*, where items occur in sequence one after another, and (b) *dictionaries of items*, where every item is associated with a unique key (or label) that

```
row no,id,created at ,from user id ,from user ,to user id ,to user ,language,source,text
1,1.22E+18,Wed Feb 05 05:51:33 IST 2020,The Economist,5988062,?,-1,en,SocialFlow,"Globally, road accidents kill more people every year than malaria or HIV/AIDS https://t.co/sXJK9Jy7D2"
2,1.22E+18,Tue Feb 04 07:30:14 IST 2020,QuickTake by Bloomberg,2527510061,?,-1,en,SocialFlow,Drones on the island of Zanzibar are spraying a silicone-based liquid to help slow the spread of malaria https://t.co/fUvg680Qg7
3,1.22E+18,Tue Feb 04 23:26:39 IST 2020,Dr. Joachim Mabula,41779021,?,-1,en,Twitter for Android,"Did you know that #Cancer was responsible for more deaths in 2018 than HIV/AIDS, Malaria and TB combined? Early detection is the key to tackling cancer globally. https://t.co/1k9UhphHVX #WorldCancerDay https://t.co/2BfUAyMyuA"
4,1.22E+18,Wed Feb 05 10:56:41 IST 2020,Ashiambi Wa Ndukwe,2460752279,?,-1,en,Twitter for iPhone,"RT @Asamoh_: Moi was MP for 45 years. He was president for 24 years. Baringo residents are still food insecure , dying of malaria and eati…"
5,1.22E+18,Wed Feb 05 10:56:06 IST 2020,Eightball85Vb5Vb5Vd5Vd5,1.19E+18,realDonaldTrump,25073877,en,Twitter for Android,"@realDonaldTrump Nice game show! Malaria, show the contestant what he won! Ooohh, a beautiful star on a navy blue ribbon! And you little girl, what door do you choose? Ooohh, a beautiful Opportunity Scholarship! What a guy! õY™„õY™…õY™™"
6,1.22E+18,Wed Feb 05 10:55:42 IST 2020,Mummy's Girl õY™„ã€¤,1.02E+18,?,-1,en,Twitter for Android,"@AsiwajuLerry: Girls having mixed feelings, sadness, anxiety, anger, sleeping issues, malaria among others during menstrual period"
7,1.22E+18,Wed Feb 05 10:55:04 IST 2020,kristofa Tetteh Odopey,5765500826,?,-1,en,Twitter for Android,"RT @juditrius: Learning at the south-south meeting of @ADP_health, now listening to @MargaretGyapong on how the national multisectoral platâ€¦"
8,1.22E+18,Wed Feb 05 10:53:19 IST 2020,nagapuri sai charan,1440517916,?,-1,en,Twitter for Android,"We are facing health issues due to buffaloes on street we are frequently getting health issues like dengue,malaria,chikungunaya,. frequentl… We request you to take immediate action @Commissioner_GHMC @Owaisi_Online @KTRTRS https://t.co/iNO1JJTBNZ"
9,1.22E+18,Wed Feb 05 10:53:15 IST 2020,ANYONE BUT TRUMP õY˜±ñõY‡±,8.03E+18,JonathanTurley,?,-1,en,Twitter for iPhone,"@JonathanTurley Yet youâ€™re okay with the Cheeto bullying folks, name-calling, grabbing women by the p****, banging a porn star while Malaria was home with princeling, extorting a foreign country and we havenâ€™t gotten to what could be on Jeaniâ€™s dre…"
10,1.22E+18,Wed Feb 05 10:52:17 IST 2020,Dude. Sheershy #NeverTrump,4248168732,Sahilcdesaic,7.82E+17,en,Twitter for iPhone,"@Sahilcdesaic Rush von Opioidfiend is clearly closer to death than previously believed. What did Malariaã€¤â€™s green jacket day again?"
11,1.22E+18,Wed Feb 05 10:47:58 IST 2020,ArsenalTruths,9.13E+17,?,-1,en,Twitter for Android,"RT @Asamoh_: Moi was MP for 45 years. He was president for 24 years. Baringo residents are still food insecure , dying of malaria and eati…"
12,1.22E+18,Wed Feb 05 10:47:37 IST 2020,babe..,1.10E+18,?,-1,en,Twitter for Android,"RT @AsiwajuLerry: Girls having mixed feelings, sadness, anxiety, anger, sleeping issues, malaria among others during menstrual period"
13,1.22E+18,Wed Feb 05 10:45:30 IST 2020,Sandy H,3270686212,Actually3,1585117688,en,Twitter for iPad,"@actually3 @gngavon @OrgJackBrown Malaria - thatâ€™s how I refer to her, too."
14,1.22E+18,Wed Feb 05 10:44:22 IST 2020,Honeyschild,3503577700,EdanClay,7.94E+17,en,Twitter for iPhone,"@EdanClay @realDonaldTrump Malaria looks like: Oh, shit! Heâ€™s made her mad. Sheâ€™s coming at my Epstein visa. She knows..."
15,1.22E+18,Wed Feb 05 10:43:59 IST 2020,Leah,5159912573,?,-1,en,Twitter Web App,"RT @naughtywriter2: I felt like we were watching Satan's version of The Ellen DeGeneres Price is Right show starring Donald and Malaria wasâ€¦"
16,1.22E+18,Wed Feb 05 10:41:33 IST 2020,Dude. Sheershy #NeverTrump,4248168732,charlieki411,29292927,en,Twitter for iPhone,"@charlieki411 @DonaldJTrumpJr She was prepping trumpâ€™s toilet paper, as requested by Malaria. He turned around and gave his daughter wife a big smile and ignored Malaria"
17,1.22E+18,Wed Feb 05 10:41:14 IST 2020,Wavyson,1.03E+18,?,-1,en,Twitter for Android,"@bexxta_: omg this reminds me of the time i told my neighbor's cousin about rice & beans underbridge with his sugar daddy on his stomache…"
18,1.22E+18,Wed Feb 05 10:40:22 IST 2020,Good Witch of Emma,3503371379,?,-1,en,Twitter for Android,"RT @SpeakerPelosi: Malaria looks like: Oh, shit! Heâ€™s made her mad. Sheâ€™s coming at my Epstein visa. She knows..."
19,1.22E+18,Wed Feb 05 10:40:00 IST 2020,Leah,5159912573,?,-1,en,Twitter Web App,"RT @naughtywriter2: I felt like we were watching Satan's version of The Ellen DeGeneres Price is Right show starring Donald and Malaria wasâ€¦"
20,1.22E+18,Wed Feb 05 10:39:18 IST 2020,Dude. Sheershy #NeverTrump,4248168732,charlieki411,29292927,en,Twitter for iPhone,"@charlieki411 @DonaldJTrumpJr When pple eat live frog õY, and all sorts of creatures, what were they expecting? Malaria?"
21,1.22E+18,Wed Feb 05 10:38:27 IST 2020,Mazi Peter Olisa,1571795894,chukudebalu,2668067464,en,Twitter for iPhone,"@chukudebalu When pple eat live frog õY, and all sorts of creatures, what were they expecting? Malaria?"
22,1.22E+18,Wed Feb 05 10:37:16 IST 2020,Myandra is Mesmeric ã/,Graceless Electric Lady,5995582707,?,-1,en,Twitter Web App,"I felt like we were watching Satan's version of The Ellen DeGeneres Price is Right show starring Donald and Malaria was the Jester turner. Byoulie"
23,1.22E+18,Wed Feb 05 10:37:15 IST 2020,Dr. Kâ€¡âˆžLOO,2767438544,?,-1,en,Twitter for iPhone,"RT @AsiwajuLerry: Girls having mixed feelings, sadness, anxiety, anger, sleeping issues, malaria among others during menstrual period"
24,1.22E+18,Wed Feb 05 10:36:57 IST 2020,danyola õYš,8.98E+17,?,-1,en,Twitter for Android,"RT @MelanieEmmajade: Malaria is complicated and spread by mosquitoes. @rihickson summing up why modelling malaria is challenging and requiâ€¦"
25,1.22E+18,Wed Feb 05 10:36:56 IST 2020,Somya Mehra,9.13E+17,?,-1,en,Twitter for Android,"RT @MelanieEmmajade: Malaria is complicated and spread by mosquitoes. @rihickson summing up why modelling malaria is challenging and requiâ€¦"
26,1.22E+18,Wed Feb 05 10:36:05 IST 2020,Favour,8.19E+17,?,-1,en,Twitter for Android,"RT @AsiwajuLerry: Girls having mixed feelings, sadness, anxiety, anger, sleeping issues, malaria among others during menstrual period"
27,1.22E+18,Wed Feb 05 10:35:15 IST 2020,Sumit Gupta,609954154,?,-1,en,Twitter for Android,"RT @TheSanjivKapoor: Some perspective: In the past week alone, while India fretted about Coronavirus,~3500 people have been killed in roadâ€¦"
```

Fig. 2.4 Example of tweets for malaria query in CSV format

Fig. 2.5 Example of JSON
format

```
                  {
    "firstName":"Kirat",
    "lastName":"Sharma",
    "age":22,
    "address":{
        "streetAddress":"22 4th Street",
        "city":"Seattle",
        "state":"WA",
        "postalCode":"98116"

},
    "phoneNumber":[
        {
            "type":"home",
            "number":"312 888-6656"

},
        {

            "type":"fax",
            "number":"454 999-9876"

}

],
      "gender":{
        "type":"male"

}
      }
```

identifies or names that item. In each case, an item can either be a string (a piece of
text), a list, or a dictionary. This recursive aspect of the JSON format can represent
data with more complex structures than the single table of information described in
a CSV file. In a JSON file (Fig. 2.5), (i) strings and keys are enclosed in pairs of
quotation marks; (ii) lists are enclosed between the symbols [and], with a comma
between each item; and (iii) dictionaries are enclosed between the symbols { and },
with each entry shown in the format *key:value*.

2.3.3.1 Typical Uses

JSON is widely used as a lightweight but flexible mechanism for storing and
exchanging structured data between different systems, especially in web-based
application programming interfaces (APIs). It facilitates communication and data
sharing between independent online platforms.

2.3.3.2   Advantages

1. It can be quickly processed programmatically.
2. Despite defining only a minimal set of structures, these structures are sufficient to represent complex data when used together.
3. It provides an effective way of representing structured data that can be quickly processed by computer software.

2.3.3.3   Disadvantages

Strings are atomic and can have no further structure in JSON, making the format less suited to represent additional information about parts of documents that are primarily textual.

2.3.4   XML

eXtensible Markup Language (XML) is a format with many similarities to HTML and is used to represent complex documents in a way that can be reliably processed automatically. XML is different from HTML, as it primarily specifies the syntax of a language (Fig. 2.6). This means that tags begin with the form <tagtype> and end in the form </tagtype>, whereas HTML additionally specifies a vocabulary and semantics. The syntax of HTML and XML are very similar; XML, though, imposes stronger formal requirements on a document. Unlike HTML, which is a continually evolving standard with new elements being added and revised over time, XML

Fig. 2.6 Example of an XML format

```
<title>
    <author>Jack</author>
    <published='years'>3</published>
    <type>book</type>
    <isbn>9782012128011</isbn>
</title>
    <title>
    <author>Toby</author>
    <published='years'>10</published>
    <type>manuscript</type>
    </title>
<title>
    <author>Sugar</author>
<published='years'>3</published>
    <type>video</type>
</title>
</title>
```

defines syntactical rules. The majority of software processing XML is designed to confirm the validity of an XML document according to this predefined grammar before processing it and to reject (i.e., refuse to process) invalid documents.

2.3.4.1   Typical Uses

XML is better suited than HTML to represent documents that should be processed by automated software for purposes where more precision is needed than in the case of merely displaying information—a task in which some degree of error is often tolerable. As a result, XML can be used when HTML is unsuitable, such as storing program configuration data or, like JSON, exchanging information via API. In addition, XML is often used in the humanities for annotating textual materials with additional information that can subsequently be processed programmatically.

2.3.4.2   Advantages

1. As XML is a markup language, it is well-suited in layering additional information on top of what is essentially a single coherent textual document. Furthermore, since XML itself describes only the grammar used to convey this information and not the vocabulary and semantics, additional standards are used in conjunction with XML to facilitate the interchange of documents.
2. Because XML documents must be valid to be processed, unlike an HTML document, an XML document always has one unique correct interpretation.

2.3.4.3   Disadvantages

1. Like HTML, XML imposes a strict hierarchy, where every element must be entirely enclosed by some single more significant element, which causes difficulties when representing information about overlapping textual content.
2. XML documents are often complex and can be harder to read than simpler formats, such as JSON.

2.3.5   Binary Files

File types can be separated into two classes: text-based formats and binary (i.e., non-text-based) formats. Text-based formats represent their contents on the most basic level as a sequence of characters, whereas binary formats typically use a mixture of representations for storing data. When viewed in a text editor, the contents of any text-based format file will appear as a coherent mixture of characters and a relatively small number of codes (Fig. 2.7). In contrast, any file in a binary format,

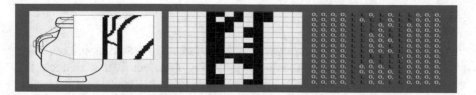

Fig. 2.7 Example of binary format

when opened in a text editor, will typically display large amounts of garbled, unintelligible material, if it can be opened at all. Binary files represent their contents using methods other than character encodings. For example, a text file containing the content "123" will consist of numbers representing a sequence of characters "1," then "2," and then "3," using an encoding such as UTF-8, resulting in a series of numbers, each representing one of these three characters. In contrast, a binary file will instead often represent the numeric value 123 directly as a number. Because this representation is different from any character encoding, a text editor typically cannot display the content in a meaningful way regardless of which character encoding is used. A binary file generally requires special software designed to process the specific format of a binary file.

2.3.5.1   Typical Uses

Because of their efficiency in storage space and processing speed, binary files are often used to store data types that result in large files, such as images, video, audio, executable code, and compressed files.

2.3.5.2   Advantages

1. They use storage space efficiently.
2. They are often easier for software to process, as their contents frequently correspond more closely to data structures used internally by software for processing them.

2.3.5.3   Disadvantages

1. They are not human-readable, and their contents cannot be processed or interpreted without precise and detailed knowledge of their format.
2. It is not generally possible to usefully display or edit the contents of a binary file without knowing its exact format.

2.4   Metadata

Metadata provides *data about data* or *information about information*. Data might be useless and considered incomplete if lacking metadata. It is a valuable tool that tells about the details of an author, place, title, and other unique resource identifiers. Thus, it helps to describe the quality, content, condition, and other dataset characteristics. In a traditional library, a catalog that describes library holdings is the metadata, whereas, in a digital environment, metadata is the search term described by someone in a text form. We find new ways with artificial intelligence and other computational forms to determine new kinds of search methods with time. The various functions of metadata include the following:

- Discovery of resources
- Organization of electronic resources
- Providing digital identification
- Archiving and preservation of resources
- Accessing the resources
- Minimizing the duplication of data
- Fostering the sharing of digital data resources
- Preserving the efficacy of data over time by describing the various methods used for data collection and dataset creation during a research process
- Capturing the information for data discovery
- Reusing the scientific data (reproducibility)
- Defending the science behind decisions (i.e., if challenged in court, metadata can help to take appropriate decisions)
- Increasing the visibility of the data when used for citation purposes
- Data management, data distribution, and project management
- Tracking data provenance accurately (to determine if the data is in current or in a reliable format)
- Data liability
- Ensuring the institution's investment in the data by possessing information about how the data was processed, quality controlled, and collected
- Making a permanent record of the dataset
- Creating new impending partnerships
- Promoting the organization's research
- Collaborating through data sharing

Thus, a metadata record is a declaration of its (a) purpose, (b) use constraints, (c) completeness, and (d) distribution liability. Many data portals consist of accessible metadata collections from different organizations. Some of the examples of metadata search catalogs are shown in Table 2.3.

Table 2.3 Examples of metadata repositories

Data portal	Description
DataONE Search https://search.dataone.org/data	Metadata repository, available to all researchers.
Data.gov http://www.geo.data.gov	Federal e-gov geospatial data portal
Metacat http://knb.ecoinformatics.org/index.jsp	Repository for data and metadata
US Geological Survey http://data.usgs.gov/datacatalog	USGS Science Data Catalog
ArcGIS Online http://www.geographynetwork.com	ESRI-sponsored national geospatial data portal

2.4.1 What Is a Metadata Standard?

A metadata standard provides common terms, definitions, and structure to describe data with consistency, more straightforward interpretation, ease of communication, and easy retrieval (Table 2.4). It presents a uniform summary description of a dataset. It is made up of defined elements. Its predictable format helps in fast computer interpretation and processing and supports search and retrieval in automated systems. Many tools are available to make metadata like Morpho, XML Spy, CatMD, or EPA Metadata Editor.

Table 2.4 Examples of common metadata standards

Metadata standard	Type of data
Dublin Core Element Set http://dublincore.org/documents/dces/	Web resources, publications.
FGDC Content Standard for Digital Geospatial Metadata (CSDGM) https://www.fgdc.gov/metadata/csdgm-standard	Data portal geospatial data
ISO 19115/19139 Geographic information metadata https://www.fgdc.gov/metadata/iso-standards	Geospatial data and services
Ecological Metadata Language (EML) http://knb.ecoinformatics.org/eml_metadata_guide.html	Ecological data
Darwin Core http://rs.tdwg.org/dwc/index.htm	Museum specimens
Geography Markup Language (GML) http://www.opengeospatial.org/standards/gml	Geographic features (roads, highways, bridges)

2.4.2   Steps to Create Quality Metadata

The common steps for creating a quality metadata are:

1. Organize one's data.
2. Create metadata employing a metadata tool.
3. Check the accuracy and completeness of the created metadata.
4. Check the metadata record and revise accordingly.
5. Review the record before one publishes.
6. Do not use tabs, indents, and special characters while creating one's metadata.
7. Use a text editor to prevent adding unnecessary characters in the background of the text when copying content from other sources.

Metadata is as valuable as data. Also, to summarize the details about print resources, metadata can summarize details about sites, web pages, images, videos, blogs, and tweets. After querying in a database, metadata is often displayed as search results and usually expressed in the form of tags. Metatags are regularly evaluated to decide the position of the search results and their relevance when indexing various resources in a database. They can be created manually or by automated information processing (see Chap. 4). Manually created tags are more accurate than automated tags and are constructed according to the creator's knowledge and according to his/her interpretation of the information. Till now, there is no standard for sharing and reuse of metadata for many repositories (Table 2.5). Further, metadata summarizes necessary information about data and, thus, is used popularly in text and data mining to make sense from a large corpus of data. The analyzed data from metadata can link various resources and help develop better semantically linked resources available on the library site or other resources.

Table 2.5 Common concerns and solutions regarding metadata

Common concerns	Solution
How to capture accurate, robust metadata?	It is suggested to incorporate metadata creation in the data development process in a distributive effort
How to invest time and resources to create, manage, and maintain metadata?	It is suggested to be included in the grant budget and schedule
How to ensure the readability/usability of metadata?	The best practice is to use a standardized metadata format
How to use discipline-specific information and ontologies?	It is suggested to use a standard profile that supports discipline-specific information

2.5 Digital Data Creation

Library data is often generated from physical or analog sources. Digitizing analog text data can be challenging as it involves transforming anything from handwritten to printed text into digital text representations. Typically, library practitioners will use a computational strategy called optical character recognition (OCR) to recognize text characters in images taken of a text automatically. It can be a quick and effective approach for creating digital text data from an analog textual source without transcribing the text data manually. OCR outputs can be used to create datasets for text analysis. The following are a few examples of text analysis projects that used OCR: *National Security Archive's Kissinger Collections*,[1] *The Dispatch*,[2] and *Women Writer's Project*.[3] Text analyses are conducted to explore vast collections of information and perhaps to create new understandings of those materials and their significance, using methods that humans cannot conceivably complete in a reasonable amount of time without computational support. Some of the popular tools for OCR include ABBYY FineReader,[4] Acrobat Adobe,[5] Amazon Textract,[6] Google Cloud Vision,[7] Tesseract.[8] and Pytesseract.[9] In order to perform OCR on a text corpus, one needs to do the following (Fig. 2.8):

1. Prepare a single file folder containing all of the corpus files. If the corpus is small enough (e.g., one book), this could be simply a single file (e.g., a .pdf).
2. All corpus files should be of the same file format.

Scanned Document   Image or PDF File   OCR   Text File

Fig. 2.8 Workflow of optical character recognition (OCR)

---

[1] https://blog.quantifyingkissinger.com/.

[2] https://dsl.richmond.edu/dispatch/.

[3] https://wwp.northeastern.edu/.

[4] https://pdf.abbyy.com/.

[5] https://acrobat.adobe.com/us/en/acrobat.html.

[6] https://aws.amazon.com/textract/resources/?blog-posts-cards.sort-by=item.additionalFields.createdDate&blog-posts-cards.sort-order=desc.

[7] https://cloud.google.com/vision/docs.

[8] https://tesseract-ocr.github.io/tessdoc/Home.html.

[9] https://pypi.org/project/pytesseract/.

3. The chosen file format should be interoperable and stable.
4. Create an organized file structure so that the file naming structure identifies both the volume of the images and their corresponding scanned page. It helps to maintain the order of the volume (e.g., one file folder for the entire corpus and one sub-folder for each volume in the corpus containing an image file for every page in the volume).
5. From the images, the OCR process will create plain text.
6. Organize a corpus of such text in one file folder containing one file per volume in the `.txt` (plain text) format. The plain text file format is interoperable, stable, and fully computer-readable, meaning it will be ready for performing computational analysis and for storing in repositories and databases.

However, in some instances, OCR does not work, and library practitioners may need to manually transcribe some of the text data. In addition, OCR does not necessarily guarantee that the text will be interpretable by the computer. Thus, additional steps, such as manually identifying words and characters, may need to occur if OCR is unsuccessful. In practice, handwritten text, text typed by typewriter, and scanned files without OCR will produce more quirky characters and will not be successfully recognized by the computer. To overcome the above problem, many libraries, including the Library of Congress, Smithsonian Transcription Center, and British Library, use crowdsourcing to digitize textual information. Crowdsourcing refers to using online volunteers (or minimally paid, in services like Amazon's Mechanical Turk) to manually transcribe texts or perform other tasks. Transcription crowdsourcing distributes the task of transcribing a corpus of texts among a large group of people (potentially worldwide) and makes it possible to manually digitize more material than would be possible by the original project team alone.

Alex Wermer-Colan works as the Digital Scholarship Coordinator at Temple University Libraries' Loretta C. Duckworth Scholars Studio, formerly known as the Digital Scholarship Center where Alex held a Council on Library and Information Resources Postdoctoral Fellowship. His Ph.D. in English involved a specialization in critical theory and natural language processing with a focus on twentieth-century Euro-American culture. At Temple Libraries, he has collaborated with librarians, archivists, students, faculty on an ongoing digitization project of mass-market science fiction. Alex regularly advises and collaborates with faculty and students on text mining projects for curating library collections as data, conducting research in fields of cultural analytics and web studies, and training graduate students in emerging methods in interdisciplinary research and teaching.

Story: The SF Nexus
The vast majority of twentieth-century literature remains undigitized, unavailable for access online even for the purposes of data analysis. The field of "distant reading" in literary criticism remains dependent upon

(continued)

impoverished datasets that are not representative of book history, especially mass-market genre fiction. We began a digitization project of mass-market science fiction at Temple University Libraries in the fall of 2017. We are currently expanding the project to include a range of special collections around the country with science fiction holdings.

Because copyright restricts access to most twentieth-century literature, this project is inspired by the collections as data movement to make the literary works available in a disaggregated form to researchers and students. We have received regular requests for access, and data curation is often customized for individual researchers. The workflow is iterative, and as the corpus grows, its value to researchers will increase as well. We are ingesting all the works into HathiTrust Digital Library for access through their Research Center. We also intend to evaluate copyright and rights holders of individual items to identify which can be displayed online. Furthermore, we are developing data-sharing agreements to enable libraries to exchange copyrighted corpora.

This project focuses on curating textual data (and images) for a wide range of text mining methods. We have extracted features at the chapter level as bags of words to enable topic modeling—one master's thesis has already been completed using the science fiction corpus as disaggregated data. We are also conducting research on the corpus—we received a grant through Temple University to lead a project on climate fiction, entitled "The Stories We Tell Ourselves: Cultural Analytics of Climate Fiction in the Age of the Anthropocene." While we will use elementary methods of text mining to count words and collocations, my interest lies in recent innovations in machine learning, especially vector space modeling or word embeddings, for understanding how words relate to one another in a text or corpus. While I have used a wide range of user-friendly tools to teach text mining, such as Voyant Tools or Ubiquity's DocuScope, most of our text mining methods depend on R and Python libraries, such as Gensim for topic modeling and Word2Vec.

This project makes available a neglected corpus of literary history in a multimodal format. We aim to create a website that will showcase user-friendly forms of access while teaching more advanced modes of text mining using extracted features sets. The difficulty to access is primarily caused by the copyrighted nature of the works. Working with extracted features sets is, in some ways, more difficult than using full texts. Although if curated in a pedagogical way, extracted features may be a more ripe opportunity for teaching the fundamentals of text mining and the limitations to various forms of processing. The main barriers are the contingent nature of many staff members' time on projects. This project has depended on graduate students with limited contracts, and the project's focus for this reason has been as

(continued)

much on the pedagogical value of teaching these methods and developing the corpora. The other major obstacle is copyright and the bias in libraries towards digitizing works that can be exhibited as facsimile without risk of lawsuits. The project is still growing, but maintaining the quality of the scanning process and curating the data for researchers can be taxing and unpredictable.

Supplementary Media Links

1. https://sites.temple.edu/tudsc/2017/12/20/building-new-wave-science-fiction-corpus/
2. https://sites.temple.edu/tudsc/2018/04/26/modeling-the-new-wave-on-learning-to-use-machines-to-read-sci-fi-lit/
3. https://sites.temple.edu/tudsc/2019/07/18/curating-copyrighted-corpora-an-r-script-for-extracting-and-structuring-textual-features/
4. https://news.temple.edu/news/2020-06-15/byte-size-books-digitizing-temple-university-libraries-science-fiction-collection
5. https://lcdssgeo.com/omeka-s/s/scifi/page/digitizing-science-fiction

2.6 Different Ways of Getting Data

There are various ways to get data. We will explore several of the most common approaches, for instance, how to identify online data repositories and download data from them, how you might access data stored in relational databases, as well as data accessible via web APIs. For those cases where online data is not readily available via APIs or repository downloads, we additionally discuss the use of digital data and web scraping to get data from those sources.

Amy J. Kirchhoff is the Constellate Business Manager and Archive Service Product Manager for Portico, JSTOR.

Story—Constellate: Text Analytics Platform
JSTOR and Portico are developing a text and data analytics service that provides "a platform for learning and performing text analysis, building datasets, and sharing analytics course materials" [a] (the beta release is at https://constellate.org/). This service opens the black box of text analysis: enabling librarians and faculty to teach it, students to learn it, and researchers to make discoveries using content from JSTOR, Portico, and other sources.

(continued)

Text analytics, *"or the process of deriving new information from a pattern and trend analysis of the written word, has the potential to revolutionize research across disciplines. Sadly, there is a massive hurdle facing those eager to unleash its power. All too often, researchers and students learn about the promise of text analytics, only to have it revealed that the promise can be realized solely by the select few with the necessary technical skills. Ted Underwood, Professor of English at the University of Illinois, likens this scenario to researchers being presented with"* [b] a *"deceptively gentle welcome mat, followed by a trapdoor"* [c]. *To help solve this problem, the JSTOR and Portico text analytics service "is centered on student and researcher success, providing text and data analysis capabilities and access to content from some of the world's most respected databases in an open environment with access to a variety of teaching materials that can be used, modified, and shared."* [a]. *The platform has three pillars:*

1. *Collections and Dataset Builder: This currently includes analytics access to anchor collections from JSTOR and Portico, with additional collections (such as the Library of Congress digitized newspapers in Chronicling America). Users have the ability to download datasets of bibliographic metadata, unigrams, bigrams, trigrams, and full text (depending upon the rights associated with the content). Users have access to a dashboard to view datasets they have previously created or accessed.*
2. *Visualize and Analyze: This currently includes built-in visualizations, such as an n-gram viewer and word cloud. The platform also provides personal computational environments (individual Binder instances spun up with Jupyter and populated with tutorial and template Jupyter Notebooks for text analytics). It will eventually include a secure computing environment to allow users to work with the full text of rights restricted content.*
3. *Open Educational Resources: This currently includes template and tutorial code and lessons and educational materials for use and reuse by the community. It will eventually include collaborative resource creation where users may create, edit, resume, and collaborate in the creation of educational resources for text analysis.*

Over the course of 2020, JSTOR and Portico have worked with ten pilot institutions (https://docs.tdm-pilot.org/about/#reference-institutions) that are teaching with the tool, doing research on the platform, and building Python tutorials. The platform will be released in 2021 with three tiers of service, starting at free and then with two participation packages where fees are tiered to institutional size, providing both larger dataset sizes and computational resource allocations at the higher tiers.

(continued)

References

[a] https://docs.constellate.org/about/
[b] https://labs.jstor.org/projects/text-mining/
[c] https://tedunderwood.com/2018/01/04/a-broader-purpose/

2.6.1   Downloading Digital Data

Various open-source platforms and tools help to retrieve digital trace data (see Sect. 2.1.1), such as tweets, Wikipedia pages, YouTube comments, Facebook posts, articles, and newspapers by using the API(s) provided by the databases. Some of the popular platforms that offer free access to digital trace data are summarized in Table 2.6.

Moreover, various libraries in R and Python can retrieve digital trace data from various databases and sources with specific API restrictions on the download (Table 2.7). These restrictions are elaborated in the API documentation of the respective databases.

2.6.2   Downloading Data from Online Repositories

A great deal of data is available for download as pre-structured files through online repositories. This data is commonly created and hosted by governments, universities, or other large institutions. Getting data from these repositories is often as simple as selecting the datasets one wants through the repository's user interface by clicking "download." Table A.1 presents selected online repositories available for text mining. Other ways of extracting data are through web APIs and web scraping.

2.6.3   Downloading Data from Relational Databases

A database is usually an organized collection of data in a digital system that allows for its creation, modification, and retrieval. It uses queries to request data matching by specifying a set of formal requirements that makes it possible to ask precise and unambiguous questions about aspects of all of the data contained in the database in a way that database software can respond effectively. The most common type of database is a relational database, in which data is structured and organized conceptually into one or more tables. Many databases subscribed by libraries or library management systems (LMS) used by libraries are in the form of relational

Table 2.6 Example of tools available for collecting digital trace data

Platform/tool	Description	Data	Access	Limit
Digital Scholar Workbench—Constellate https://constellate.org/	Provides access to CSV containing only metadata and JSON file containing metadata and textual data that includes unigrams, bigrams, trigrams, and full text (where available) for text analysis	Articles, books, chapters, and newspapers from JSTOR, Portico, Chronicling America, CORD-19, and DocSouth databases	Free	Up to 25,000 documents in a dataset where a single IP can create 10 datasets per day
TDM Studio https://tdmstudio. proquest.com/home	Provide access to ProQuest's collection across all disciplines and formats. Researchers can also incorporate their own datasets such as institutional repositories and journal articles and make use of additional content such as social media and blogs. The platform further allows researchers to perform text mining and visualization techniques	Current and historical newspapers, full-text dissertations and theses, scholarly journals, and primary sources	Subscription-based	10 datasets where one dataset limits to half a million records with a maximum limit of 15MB export of code and resultant dataset per week
Netlytic https://netlytic.org/	Cloud-based tool that uses public API to collect data and also provide basic text analysis and visualization	Twitter, Instagram, YouTube, Facebook	Free access up to 2500 records but paid if want data up to 100,000	1000 most recent tweets
Twitter Archiving Google Sheet (TAGS) https://tags.hawksey.info/	Google spreadsheet template that allows to download tweets for different hashtags automatically	Twitter	Free	API limit

(continued)

Table 2.6 (continued)

Platform/tool	Description	Data	Access	Limit
RapidMiner https://rapidminer.com/	Data mining tool that retrieves data from Twitter using its API and performs various data and text mining analyses	Twitter	Free	API limit
Orange https://orangedatamining.com/	Data mining tool that retrieves data from selected databases using their API and performs various data and text mining analyses	Twitter, Wikipedia, PubMed, The Guardian, NY Times	Free	API limit
Harzing's Publish or Perish https://harzing.com/resources/publish-or-perish	Standalone software that retrieves and analyzes citations from selected databases using their APIs	Google Scholar*, Crossref*, PubMed*, Microsoft Academic**, Web of Science***, Scopus***	* Free; ** Require free registration; *** Require subscription	Google Scholar has 1000 limit per query

Table 2.7 Example of various libraries from R and Python for collecting digital trace Data

Library	Description	Programming language
crossrefapi https://github.com/ fabiobatalha/crossrefapi	Retrieves data from Crossref	Python
twitter https://pypi.org/project/ twitter/	Retrieves data from Twitter	Python
tweepy https://www.tweepy.org/	Retrieves data from Twitter	Python
python-youtube https://pypi.org/project/ python-youtube/	Retrieves data from YouTube	Python
Goodreads https://pypi.org/project/ Goodreads/	Retrieves data from Goodreads	Python
arxiv https://pypi.org/project/ arxiv/	Retrieves data from arXiv preprints	Python
scihub https://github.com/zaytoun/ scihub.py	Retrieves data from Sci-Hub	Python
gtrendsR https://github.com/ PMassicotte/gtrendsR	Retrieves data from Google Trends	R
rtweet https://www.rdocumentation. org/packages/rtweet/ versions/0.4.0	Retrieves data from Twitter	R
aRxiv https://github.com/ropensci/ aRxiv	Retrieves data from arXiv preprint repository	R
rcrossref https://github.com/ropensci/ rcrossref	Retrieves data from Crossref	R
scholar https://cran.r-project.org/ web/packages/scholar/index. html	Retrieves data from Google Scholar	R
rAltmetric https://cran.r-project.org/ web/packages/rAltmetric/ README.html	Retrieves data from Altmetric database	R
bibliometrix https://bibliometrix.org/	Shiny app that retrieves metadata from PubMed, Digital Science Dimensions, and Cochrane databases using their APIs and also perform various bibliometric methods of analysis	R

(continued)

Table 2.7 (continued)

Library	Description	Programming language
guardianapi https://docs.evanodell.com/ guardianapi/	Retrieves data from The Guardian newspaper	R
dimensionsR https://cran.r-project.org/ web/packages/dimensionsR/ index.html	Retrieves data from Dimensions database	R
rgoodreads https://github.com/Famguy/ rgoodreads	Retrieves data from Goodreads	R
gutenbergr https://github.com/ropensci/ gutenbergr	Retrieves data from Project Gutenberg	R
wosr https://cran.r-project.org/ web/packages/wosr/wosr.pdf	Retrieves data from Web of Science database	R
vosonSML https://github.com/vosonlab/ vosonSML	Retrieves data from Reddit, YouTube, Twitter,	R
rscopus https://cran.r-project.org/ web/packages/rscopus/ rscopus.pdf	Retrieves data from Scopus database	R

databases. Each table of a relational database consists of named columns, each representing some particular attribute, and rows, each representing a relationship between specific values of those attributes. Each row often corresponds to a concrete entity, such as a person, place, object, or an abstract entity, such as a relationship between two objects. In relational databases, Structured Query Language (SQL) is usually used to allow the precise formulation of queries and statements describing the creation and modification of data and tables in a database. For example, the following relational database describes a list of library holdings, a list of thematic categories, and the relationships between the two. Each book has various properties, such as title, author, and year of publication, and each book may belong to any number of categories. Each category may similarly apply to any number of books. The database consists of three tables of data:

1. A "book" table, which contains information about one book in each row (Table 2.8).
2. A "category" table, which contains information about one category in each row (Table 2.9).
3. A "hascategory" table, in which each row represents the fact that a particular book belongs in a particular category (Table 2.10).

Table 2.8 Example of a *book* table

Book_id	Hollis_id	Title	Author	Publish_location	Publisher	Publish_date	Language
1	000000130	Bone Crier's Moon	Kathryn Purdie	New York	Katherine Tagen Books	2020	English
2	000000277	The Dragon Egg Princess	Ellen Oh	New York	Harper Collins	2020	English
3	000000296	Revolver Road	Christi Daugherty	United Kingdom	Minotaur Books	2020	English
4	000002169	Do Androids Dream of Electric Sheep?	Philip K. Dick	United States	Ballantine Books	1996	English
5	000002856	Victoria and Abdul: The True Story of the Queen's Closest Confidant	Shrabani Basu	United Kingdom	History Press	2010	English
6	000005878	The Vanishing Deep	Astrid Scholte	United States	G.P Putnam's Sons Books for Young Readers	2020	English

Table 2.9 Example of a *category* table

Category_id	Description
1	Fiction
2	Fantasy
3	Sci-Fi
4	Biography
5	Thriller

Table 2.10 Example of a *hascategory* table

Book_id	Category_id
1	2
2	2
3	5
4	3
5	4
6	5

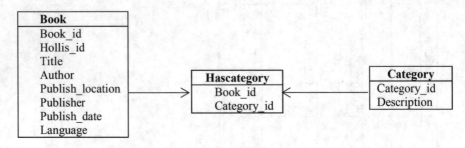

Fig. 2.9 Example showing a logical relationship between different tables in a relational database

Figure 2.9 represents how the three tables can be related logically through their shared use of identifiers for individual books (the numbers in the Book_id columns) and distinct categories (the numbers in the Category_id columns).

Representing the same information using a single table of data would be impossible without repeating the same book in multiple rows. It is, therefore, better to use numerous columns to represent all the categories that one book can belong to or record all of the categories in a single column. Figure 2.9 shows an example of a simple dataset that can be most naturally modeled using more than one table.

2.6.3.1   Use Case

Databases are frequently used in cases where structured data should be stored and queried. This includes research databases, in which the data being modeled is of direct interest to researchers, and many indirect use cases, in which the database itself may be instrumental to some other end purpose. Examples of the former are the China Biographical Database and Ming-Qing Women Writing's Project, both of which, in addition to providing different methods of access, make their data downloadable directly as an SQL database for research use. Examples of the latter include the majority of modern sites and content management systems (e.g., Wikipedia, WordPress, etc.), many of which are implemented using software that stores content in a relational database system for effectively and efficiently managing and retrieving it.

In many cases, researchers may not have direct access to the relational database (particularly those created by commercial providers) itself, except through the special permission of the database provider. Also, the interface may not expose all data directly corresponding to the underlying representation of the data in tables. In these cases, the line between a relational database and a general database or repository becomes less distinct. It may not be transparent or even relevant to the user whether the system is implemented as a relational database if the relations are not directly accessible. Examples of this approach include Operabase[10] (a database of opera performances around the world) and the Getty Provenance Index[11,12] (a database of provenance information about artistic works).

2.6.4   Web APIs

Sites provide access to online information and services in a form suitable for inter-active display in a web browser software that takes a semi-structured representation of various types of content, for example, formatted text, video, image, or audio, and turns this into something that a user can easily view, understand, and respond. The information transmitted to the web browser is semi-structured and contains some machine-readable information, such as formatting instructions or references to other web pages. However, the structured elements are typically closely associated with the *presentation* of the content, how it should be displayed in a web browser rather than its *semantics*, i.e., the logical structure of the information being shown. It means that even though the software can easily display the contents of a web page, it is much harder for it to extract useful information from it and perform any further processing.

A web-based application programming interface (API) is a system based upon the same technical foundations as popular sites but intended for computer software to interact with rather than directly accessed by a user via a web browser. Instead of providing data in HTML (HyperText Markup Language, a format for describing the content in a web page), APIs' data is usually offered in a semi-structured format such as JSON or XML containing semantic markup only. Rather than indicating how different parts of the content should be displayed, these documents describe relationships between different elements within it, for example, associating

---

[10] https://www.operabase.com/en.

[11] https://www.getty.edu/research/tools/provenance/search.html.

[12] https://piprod.getty.edu/starweb/pi/servlet.starweb?path=pi/pi.web.

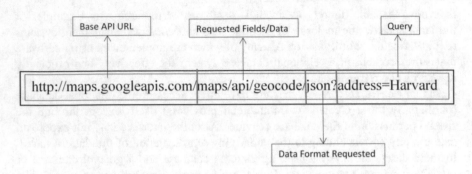

Fig. 2.10 Anatomy of an API call

different properties as belonging to some particular entity. Instead of using methods
like hyperlinks and search boxes for navigation (as in HTML), web APIs rely
upon documented ways of constructing URLs (Universal Resource Locators) and
formally specified conventions for content submission, which may be specific to the
particular API in question. APIs usually regulate (i) who gets to access the content,
(ii) what kind of information can be obtained, and (iii) how often one can access it.
It sets rules for what one can do with API. So, when we are constructing API calls,
we are looking for the generic structure of a URL and then switching the term we
want to encode. For instance, Fig. 2.10 can be used to extract latitude and longitude
for Harvard University. We can switch from Harvard to other universities and get
the latitude and longitude by parsing JSON data.

Further, Chinese Text Project API[13,14] is an example of a web API used in a
digital library, which provides access to textual data and metadata for text mining
and integration with other online projects. Publicly available API functions listed
on the documentation page contain example links showing the JSON responses
which an API produces in response to actual requests. These can be viewed in a web
browser to help developers understand the API structure when creating applications.
Other online tools, such as Text Tools[15] and MARKUS,[16] use API to import text and
structured data from the library.

Breeding [5] in his paper emphasized that "APIs provide an essential technical
mechanism for integration, interoperability, and the extension of functionality. Many
possibilities remain within the library environment where specialized content or
service activities currently delivered through disparate interfaces could potentially
be brought together in a more user-friendly way via integration into the prevailing
interface of the library's web presence." He further highlighted that "with the
increasing availability of APIs, libraries no longer have to be satisfied with using

---

[13] https://ctext.org/.

[14] https://ctext.org/tools/api.

[15] http://ctext.org/plugins/texttools/#help.

[16] https://dh.chinese-empires.eu/markus/beta/.

their sites to deliver their users to external service providers. Instead, there are increasing opportunities to create a more sophisticated and integrated web presence able to fulfill patrons' information needs while strengthening awareness of and engagement with the library" [5]. Some of the ways by which APIs can be used in libraries are by the following:

1. Adding geocode to library OPAC by adding latitude and longitude to the name of a place, city, or country using Google API and enriching authority data in MARC21
2. Determining the satisfaction level of users for library products and services by performing sentiment analysis of comments or posts or tweets after fetching them from Twitter or Facebook or other social media platform of interest using their APIs
3. Determining the popularity of the book and deciding whether to include it in the library collection or not by performing sentiment analysis of reviews on books from Goodreads or Google Books after fetching them from their APIs
4. Enriching the metadata for library resources from different libraries' APIs like Harvard, OCLC, publisher's sites like Springer or Elsevier, or Library of Congress
5. Adding information, images, tables, or diagrams to a digital exhibition of library resources using API from Wikipedia or other open-source resources
6. Classifying the genre of videos, books, DVDs, CDs, and other materials based on identifiers like ISBN or OCLC Number using Goodreads or OCLC APIs
7. Searching users by barcode and retrieve basic shared metadata of the users
8. Accessing a specific institution's license information such as License Manager
9. Linking your library resources with other e-resources by embedding API of those resources to your library site
10. Getting information about writers, authors, characters, or corporations
11. Getting vendors' details for materials, products, content, and services

2.6.4.1 Use Case

APIs provide a way to extract machine-readable data from online systems, making them well-suited to text and data mining applications. APIs are particularly advantageous for systems containing large amounts of data, which may change over time, for which it may be inconvenient or impractical to download the entire dataset in advance before making use of it. APIs can also facilitate real-time data exchange between independent projects and applications because granular requests can be made and responded rapidly. This real-time data exchange can also be used to create "mashup" systems that, from a user perspective, function as a single cohesive platform or interface but are implemented as multiple independent systems communicating through APIs to achieve this result. Mashups can allow projects to build upon the content and functionality of existing systems without having to duplicate these or incorporate complex functionality or large volumes of data directly into a single centralized system. Table A.2 enumerates selected online repositories with API available for text mining. Moreover, the case study—Sect. 7.3 in Chap. 7—retrieved tweets from Twitter's Developer app using the API credentials and performed sentiment analysis using the credentials for API for AYLIEN text mining software.

2.6.5   Web Scraping/Screen Scraping

Sites generally provide access to information in a semi-structured format called HTML (HyperText Markup Language). It is designed to convey just enough information to a web browser. It produces a graphic and interactive illustration of mixed media content with any combination of text, image, or video using a "page metaphor," where one HTML document corresponds to one arbitrarily long web page. A core goal of the structured elements of HTML is to convey enough information about how the content should be displayed and interacted with so that the client software can closely reproduce the desired layout of material on the virtual page. As a result, much of the markup code used in HTML relates closely to aspects of the design. For instance, specifying a particular piece of text should be displayed in some specific type of font or that a particular image should be located a certain distance below and to the left of some other page element. Though some elements also convey semantic information about the associated data, this is typically not the primary goal.

Nevertheless, HTML often provides the only publicly available digital interface to online content. Even though it is not intended to contain structural elements that are not designed to facilitate access to semantic units in the data, in practice, some combination of layout-related descriptors will uniquely identify a particular semantic element in a page. For example, in a newspaper site, it will frequently be the case that all headlines of a specific type are displayed in a specific typeface, size, and style and will occur in specific regular locations on each page. Though there is no way of knowing in advance what particular formatting properties these might be, they will generally be consistent within a specific site or section of a site. Once these formatting properties are known, they can extract the corresponding information from any similar page programmatically. Additionally, since most sites are database-driven, the systems implementing site functionality typically, for convenience in generating HTML content, follow precise and consistent rules. Web scraping refers to any technique using this consistency of document structure across multiple pages to extract data from them automatically (Table 2.11).

Table 2.11 Selected web scraping tools

Tool	Best for	Skill level	Export format	Cost
Octoparse https://www.octoparse.com/	Simple and dynamic websites using HTML, CSS, AJAX technologies, JavaScript, redirects, cookies, search bars, etc.	Beginner, intermediate, advanced	TXT, CSV, HTML, Excel, API (paid only)	Free and paid versions
ParseHub https://www.parsehub.com/	Simple and dynamic websites using HTML, CSS, AJAX technologies, JavaScript, redirects, cookies, search bars, etc.	Beginner	Excel, Google Sheets, CSV, JSON, API	Free and paid versions
FMiner http://www.fminer.com/	Simple and dynamic websites using HTML, CSS, AJAX technologies, JavaScript, redirects, cookies, CAPTCHA tests, etc.	Beginner	Excel, CSV, SQLite, XMLL/HTML, JSON, ODBC (Oracle, MS SQL, MySQL, etc.)	Paid versions, free trial
Outwit https://www.outwit.com/	Simple and dynamic websites using HTML, CSS, AJAX technologies, JavaScript, redirects, cookies, CAPTCHA tests, etc.; news media; social media sites; images; documents; email	Beginner	CSV, HTML, Excel, SQL databases, image files, document files	Free and paid versions
Portia https://github.com/scrapinghub/portia	Simple and dynamic websites using HTML, CSS, AJAX technologies, JavaScript, redirects, cookies, etc.	Intermediate, advanced	CSV, JSON, XML, API	Free and open-source
Import.io https://www.import.io/	Simple and dynamic websites using HTML, CSS, AJAX technologies, JavaScript, redirects, cookies, etc.	Beginner	CSV, JSON, RSS	Free and paid versions
Webhose.io https://webhose.io/	Simple and dynamic websites using HTML, CSS, AJAX technologies, JavaScript, redirects, cookies, etc.	Beginner	XML, JSON, RSS	Paid versions, free trial

(continued)

Table 2.11 (continued)

Tool	Best for	Skill level	Export format	Cost
Data Scraper https://chrome.google.com/webstore/detail/ data-scraper-easy-web-scr/ nndknepjnldbdbepjfgmncbggmopgden?hl= en-US	Simple HTML and CSS websites	Beginner	XLS, CSV, XLSX, TSV	Free and paid versions
Web scraper https://webscraper.io/	Simple and dynamic websites using HTML, CSS, AJAX technologies, JavaScript, redirects, cookies, etc.	Beginner	Free and paid, CSV, paid only, XLSX, JSON, API	Free and paid versions
Scraper https://chrome.google.com/webstore/detail/ scraper/ mbigbapnjcgaffohmbkdlecaccepngjd?hl=en	Tables on simple HTML websites	Intermediate, advanced (competency with XPATH)	Google Sheet	Free
Scrapy https://scrapy.org/	Extracts data from websites	Familiarity with Python	Any choice of format	Free
Beautiful Soup https://www.crummy.com/software/ BeautifulSoup/bs4/doc/	Parse and extracts data from HTML and XML documents	Familiarity with Python	Any choice of format	Free
Pattern https://github.com/clips/pattern	For webscraping (Google, Wikipedia, Twitter, Facebook, generic RSS, etc.), web crawling, classification (KNN, SVM, Perceptron), HTML DOM parsing, sentiment analysis, part-of-speech tagging, n-gram search, clustering, vector space modeling, graph centrality and visualization	Familiarity with Python	Any choice of format	Free
rvest https://rvest.tidyverse.org/	Scrapes (or harvests) data from web pages	Familiarity with R	Any choice of format	Free
selectr https://cran.r-project.org/web/packages/ selectr/	Translates CSS selectors to XPath expressions	Familiarity with R	Any choice of format	Free

C. Cozette Comer is a University Librarian at the Virginia Polytechnic Institute and State University, Virginia.

Story: TDM Working Group at the University Libraries at Virginia Tech
The Evidence Synthesis Librarian at the University Libraries at Virginia Tech I supports evidence synthesis projects, such as systematic reviews. These projects are resource-intensive, requiring several contributors and skillsets, typically taking 1–3 years for proper completion. Systematic reviews are intended for decision-making, so accuracy and efficiency are paramount. Thus, I am primarily interested in text and data mining (TDM) methods to improve workflow and efficiency of evidence synthesis, for example, using tools like listearchr[a] to scrape example literature for important words to include in the search strategy. I also serve as a liaison to several academic departments, where there is demonstrated need for and interest in TDM-related support. For example, I am helping a cross-disciplinary team collect and acquire access to material that will be used to improve and validate a TDM tool they have developed.

With this background, I facilitate a TDM Working Group within the libraries. The group has representation from most areas in the library, and members joined with various existing relationships with TDM; some members use TDM in their daily work, some support TDM outside of the mining itself, while others are interested in learning about TDM but do not yet support or use it. The purpose of this group is to maintain communication between library personnel about TDM activities, and in turn, enhance current activities and streamline related knowledge base(s). To this end, we are working on several projects with varied goals. Some of these projects are aimed at organizing existing knowledge. For example, we are collecting and organizing information about TDM tools currently supported by library personnel. We also maintain a TDM library guide[b]. Some projects are aimed at creating new opportunities for patrons, for example, developing a studio for mining resources in our institutional repository. We have also pursued projects for internal purposes. For example, we are mining existing contracts with vendors for language that addresses TDM.

[a] *https://elizagrames.github.io/litsearchr/*
[b] *https://guides.lib.vt.edu/tdm/home*

Web scraping is one of the primitive approaches to collecting data and can be very cumbersome and challenging to conduct. It is the process of extracting information of interest from a site of interest to create a database of interest to study a research problem. Many sites over time have implemented stringent legal restrictions on web scraping on the sites. As the technology for building sites scaled, the legal templates for specifying how people should use those sites also scaled, restricting those sites' contents that do not have any valuable or sensitive data. It is a cumbersome and frustrating process as the source code for every site is different. As you want to create a database of your interest from different queries, you must parse the source code from various sites with a different structure. Further, as sites change over time, you have to change your code also. To conduct web scraping, we first identify the web pages we can legally scrape, followed by downloading the source code, which consists of the web page's actual contents. Lastly, we parse through the source code and populate our data frame with the text/table of interest.

2.6.5.1   Use Cases

While APIs provide direct access to structured or semi-structured data, web scraping offers an indirect route to obtaining it. Where an API exists and provides direct access to the desired data, it will typically be the preferred access method. Web scraping is a form of reverse engineering, inferring from how a page is structured visually by determining the logical structure of the data. Small changes in the layout may cause subsequent web scraping to cease to work. While APIs too can change over time, changes to APIs are made with the explicit understanding that they can impact existing API users. So, provisions are usually made to manage this process and inform the users of the API. Layout changes, by contrast, will generally be assumed by site publishers to have a little significant impact on users, as these will simply result in visual differences in how pages are displayed.

Thus, the main reason for using web scraping to collect data is that no better alternative, e.g., an API or structured bulk download, exists or is accessible. In principle, web scraping is possible for almost any site because it uses aspects of data representation inherent to the web's functioning. Thus, unlike an API or data download, web scraping does not rely upon any additional actions or provisions being made by the content provider beyond making the web pages themselves available for access using a web browser. For example, Internet search engines, such as Google, rely heavily on web scraping to create an index of content available on the web. Although primarily indexing full-text content of pages, search engines in practice also attempt some degree of structured data extraction. Ranging from simply distinguishing the main contents of an individual page from "boilerplate," navigation content appears in a similar form on many pages to more detailed structures, such as publication dates and sections within pages.

Appendix A

Online Repositories Available for Text Mining

Table A.1 presents selected online repositories and Table A.2 enumerates selected online repositories with API that can be used for text mining.

Table A.1 Selected online repositories available for text mining

Repository	Description	Data types
Registry of Research Data Repositories (https://www.re3data.org)	Searchable registry of over 2000 repositories that host research data. Individual datasets may be subject to use restrictions	Archived, audiovisual, configuration, databases, images, network-based, raw, scientific, and statistical data among others
Harvard Dataverse (https://dataverse.harvard.edu)	Searchable repository of research data in a variety of formats. Individual datasets may be subject to use restrictions	Applications, audio, documents, FITS, images, tabular data, text, compressed files (e.g. ZIP)
Full-text corpus data (https://www.corpusdata.org)	Contains full-text, downloadable corpus data from six large English corpora. Individual datasets may be subject to use restrictions or require purchase	Databases, plain text
English-Corpora.org (https://corpus.byu.edu)	Contains downloadable corpora developed by Mark Davies, Brigham Young University. Individual datasets may be subject to use restrictions or require purchase	Databases, plain text
Project Gutenberg (https://www.gutenberg.org)	Offers over 58,000 free eBooks in a variety of languages	ePub, HTML, Kindle, plain text
Spatial Data Repository (https://spatialdata.dhsprogram.com/home/)	Provides geographically linked health and demographic data from DHS Program and the US Census Bureau for mapping in geographic information systems (GIS)	Various geospatial formats, CSV
Natural Earth (http://www.naturalearthdata.com)	Free vector and raster map data	ESRI shapefile, TIFF, TFW
New York University (NYU) Spatial Data Repository (https://geo.nyu.edu)	Provides a catalog of geospatial data and maps available from New York University	Image, polygon, raster, line, point, mixed

(continued)

Table A.1 (continued)

Repository	Description	Data types
HathiTrust (https://www.hathitrust.org/)	Nonprofit large-scale digital preservation repository that includes digital content from research libraries via Google Books and Internet Archive initiatives	PDF
Global NDLTD (http://search.ndltd.org/)	Open-access electronic theses and dissertations' database provided by the Networked Digital Library of Theses and Dissertations	PDF
Open Access Theses and Dissertations (https://oatd.org/)	Open-access electronic theses and dissertations database	PDF
PQDT Open (https://pqdtopen.proquest.com/search.html)	Full-text open access theses and dissertations' database	PDF
arXiv (https://arxiv.org/)	Provides open-access preprint full text in the field of physics, mathematics, computer science, quantitative biology, quantitative finance, statistics, electrical engineering and system science, and economics	PDF
biorXiv (https://www.biorxiv.org/)	Provides open-access preprint full text in the field of life sciences	PDF
Wikipedia (https://www.wikipedia.org/)	Collects and develops content for the public in an open-access environment	PDF

Table A.2 Selected online repositories with API available for text mining

Resource	Description	Fee	Result format	Limitations	Registration
arXiv https://arxiv.org/help/api/index	It provides access to both metadata and article abstracts	Free	Atom	None	None
SAO/NASA Astrophysics Data System (ADS) https://github.com/adsabs/adsabs-dev-api	It provides access to bibliographic data on astronomy and physics publications	Free	JSON	Rate limits apply	Key required
BioMed Central https://www.biomedcentral.com/getpublished/indexing-archiving-and-access-to-data	It provides access to both metadata and full-text content	Free	XML, JSON	None	Key required
Chronicling America https://chroniclingamerica.loc.gov/about/api/	It provides access to historic newspapers and select digitized newspaper pages	Free	HTML (default), JSON, Atom	None	None
Crossref https://www.crossref.org/services/metadata-delivery/rest-api/	It provides access to metadata records with Crossref DOIs	Free	JSON	None	None
Digital Public Library of America https://pro.dp.la/developers/api-codex	It provides access to metadata of its collection	Free	JSON-LD	None	Key required
HathiTrust (Bibliographic API) https://www.hathitrust.org/bib_api	It provides access to bibliographic and rights information for its collection. It does not provide API for bulk retrieval of records	Free	MARC-XML, JSON	No specific limits, however, only intended for small numbers of items Permission must be sought for bulk retrieval	None
HathiTrust (Data API) https://www.hathitrust.org/datasets	It provides access to HathiTrust and Google digitized texts of public domain works	Free	XML, JSON	No specific limits. However, consult their policies on data use	Key required

(continued)

Table A.2 (continued)

Resource	Description	Fee	Result format	Limitations	Registration
IEEE Xplore https://developer.ieee.org/getting_started	It provides metadata for the articles submitted to the database	Free	XML	Max 200 results per query	Must subscribe to or be a member of an institution that subscribes to IEEE Xplore
JSTOR Data for Research https://about.jstor.org/whats-in-jstor/text-mining-support/	It provides access to content on JSTOR for research and teaching	Free	Zip files, XML	Max 25,000 documents per dataset; users can get access to more number of datasets by special request	Requires MyJSTOR account registration
Library of Congress https://labs.loc.gov/lc-for-robots/	It provides multiple APIs available to download bibliographic data and search Library of Congress digital collections	Free	Varies	Varies	Most APIs do not require key
Nature https://dev.springernature.com/	It provides access to the metadata of its collection	Free	XML, JSON, and more	No specific limits; however, downloads should be limited to "reasonable rates" Springer Nature TDM Policy	Varies
National Library of Medicine https://eresources.nlm.nih.gov/nlm_eresources/	It provides 29 separate APIs for accessing a wide variety of content from various NLM databases	Varies	Varies	Varies	Varies
National Center for Biotechnology Information https://www.ncbi.nlm.nih.gov/home/develop/api/	It offers several public APIs to access many databases and tools, including PubMed, PMC, Gene, Nuccore, and Protein	Free	Varies	Varies	Key required for some

Organisation for Economic Co-Operation and Development (OECD) https://data.oecd.org/api/	It provides access to the top used OECD datasets	Free	JSON, XML	Max 1,000,000 results per query, max URL length of 1000 characters	None
Open Academic Graph https://www.openacademic.ai/oag/	It provides datasets for citations drawn from two large academic graphs: Microsoft Academic Graph and AMiner	Free	Zip, JSON	None	None
ORCID https://orcid.org/organizations/integrators/API	It provides researcher profile data	Free, with subscription options	HTML, XML, or JSON	Two options: 1. Users can access the free public API, which only returns data marked as "public" 2. Become an ORCID member to receive API credentials	ORCID ID Account required
Oxford English Dictionary (OED) https://developer.oxforddictionaries.com/	It provides access to its datasets	Free, with subscription options	JSON	3000 requests per month and 60 calls per minute with a free option, other options available	Key required. Academic researchers can request free access
PLoS Article-Level Metrics http://api.plos.org/alm/using-the-alm-api/	It provides article-level metrics (including usage statistics, citation counts, and social networking activity) for articles published in PLoS journals and articles added to PLoS Hubs: Biodiversity	Free	XML, JSON, CSV	Results limited to batches of 50 at a time	Key required

(continued)

Table A.2 (continued)

Resource	Description	Fee	Result format	Limitations	Registration
PLoS Search http://api.plos.org/solr/faq/	It allows PLoS content to be queried for integration with web, desktop, or mobile applications	Free	XML, JSON	Max is 7200 requests a day, 300 per hour, 10 per minute; users should wait 5 sec for each query to return results; requests should not return more than 100 rows. API users are limited to no more than five concurrent connections from a single IP address	Key required
SpringerLink http://www.springer.com/gp/rights-permissions/springer-s-text-and-data-mining-policy/29056	It provides access to the metadata of its collection	Free	XML, JSON, and more	No specific limits; however, downloads should be limited to "reasonable rates." Springer Nature TDM Policy	Varies
World Bank https://datahelpdesk.worldbank.org/knowledgebase/topics/125589	It provides access to World Bank statistical databases, indicators, projects, and loans, credits, financial statements, and other data related to financial operations	Free	Varies	Request volume limits are unspecified but should be "reasonable"	None

Adapted from ©2020 MIT Libraries—reprinted with permission. https://libraries.mit.edu/scholarly/publishing/apis-for-scholarly-resources/. Accessed 26th Feb 2020, ©2020 Purdue University—reprinted with permission. https://guides.lib.purdue.edu/c.php?g=412592. Accessed 26th Feb 2020, ©2020 USC LibGuides—reprinted with permission.https://libguides.usc.edu/contentmining/databases. Accessed 26th Feb 2020

References

1. Eagle N, Macy M, Claxton R (2010) Network diversity and economic development. Science 328:1029–1031. https://doi.org/10.1126/science.1186605
2. Xu S (2018) Issues in the interpretation of "Altmetrics" digital traces: a review. Front Res Metr Anal 3. https://doi.org/10.3389/frma.2018.00029
3. Salganik MJ (2017) Bit by bit: social research in the digital age. Princeton University Press, Princeton
4. Nicholson S (2003) The bibliomining process: data warehousing and data mining for library decision making. Inf Technol Libr 22(4):146–150
5. Breeding M (2014) The systems librarian: APIs unify library services. Comput Libr 34(3). http://www.infotoday.com/cilmag/apr14/Breeding--APIs-Unify-Library-Services.shtml. Accessed 26 Jul 2020

Additional Resources

1. https://simmons.libguides.com/c.php?g=814790&p=5814574
2. https://www.programmableweb.com/api-research
3. https://www.libcrowds.com/collection/playbills
4. https://crowd.loc.gov/
5. https://www.fgdc.gov/metadata/geospatial-metadata-tools
6. https://transcription.si.edu/about

Chapter 3
Text Pre-Processing

Abstract This chapter focuses on the theoretical framework of text data pre-processing. It describes the three levels of text representation: lexical, syntactic, and semantic. It further explains the concept of bag of words, word embedding, term frequency and weighting, named entity extraction, and parsing. The chapter is followed by a case study showing text analysis of Tolkien's books, a web project developed by Emil Johanson.

3.1 Introduction

Any text analysis process requires some form of text preparation and cleaning. It can be of basic steps, such as removing punctuation, removing URLs, or hashtags, or advanced steps, such as annotating text with parts-of-speech (POS) tags, and often depends on the type of the application and algorithm one wants to use. Text mining algorithms help understand how the text is being prepared without having a human read it, where a machine examines the individual characters, and how they are arranged. They use all the words within the corpus, be it primary keywords or the remaining general text as keywords. They do not discriminate between documents and their usage pattern of the secondary words. Not all text mining applications require the same level of text pre-processing but are domain- and task-specific. A few of the key phrases in text pre-processing are as follows:

- *Corpus*: Discursive units parsed from sources, e.g., chapters or paragraphs.
- *Chunk*: Units of your corpus is divided into articles, sections, or paragraphs.
- *Token*: Individual pieces of content (usually words) in your text.
- *Term*: Distinct set of terms (i.e., token types), often called a vocabulary or dictionary.
- *Bag of words (BoW)*: Treatment of words as single tokens by disregarding the grammar and word order.

The textual data to be fed in well-known learning methods need to be provided in a suitable format with the correct metadata for machine processes to understand (Fig. 3.1). This format is essentially a list of strings with one string per document.

M. Lamba, M. Madhusudhan, *Text Mining for Information Professionals*,
https://doi.org/10.1007/978-3-030-85085-2_3

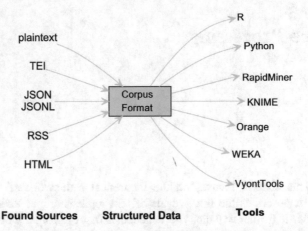

Fig. 3.1 Diagram showing conversion of textual data to structured format

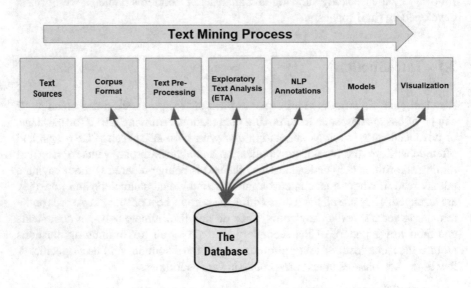

Fig. 3.2 Different stages in text mining

Each string is uniquely identified and associated with one or more labels. Once the data is transformed into this structured format, it can be fed to any programming language or tool. Figure 3.2 shows a typical text mining process where text data is first converted from its source format to a corpus format. It is then fed into a tool for pre-processing, exploratory text analysis, NLP annotations, and text mining model and is finally visualized.

3.1.1 Level of Text Representation

There are three different levels of text representation: (i) lexical, (ii) syntactic, and (iii) semantic.

1. **Lexical-Level Representation**: It is the first level of representation and consists of *words* (stopwords, stemming, lemmatization), *character* (character n-gram and sequences), *parts-of-speech tags*, and *phrases* (word n-grams, proximity features).

   a. **Character-Level Representation**: It is the first level of representation at the lexical level, where characters are categorized into letters and delimiters. They play a vital role in named entity recognition and parts-of-speech tagging. It can be applied to perform parts-of-speech tagging, spelling checking, named entity recognition, and word segmentation classification.

   b. **Word-Level Representation**: It is the second level of representation at the lexical level and is used for various text mining techniques. A *word* is the smallest delimited string of characters separated by whitespace in a sentence. Word segmentation is the problem where a word is a distinct unit in languages, such as English, but has a contrasting concept on a semantic unit where it cannot be identified by delimiters, such as in the Chinese language.

   c. **Phrase-Level Representation**: A *phrase* can be a single or group of words that work as a consequent of syntax in a sentence.

2. **Syntactic-Level Representation**: It is the second level of representation and consists of *language models*, *vector-space model*, and *full parsing*.

3. **Semantic-Level Representation**: It is the third level of representation and comprises *templates/frames*, *collaborative tagging (Web 2.0)*, and *ontologies*.

3.2 Text Transformation

3.2.1 Corpus Creation

Corpus (plural corpora) is a machine-readable collection of text that stores raw character string with metadata such as Google Books Ngram Corpus,[1] TS Corpus,[2] Nepali Text Corpus,[3] and Wikipedia Comparable Corpora.[4] A corpus can be made of a single document or multiple documents in single or multiple languages. Appendix B summarizes selected language corpora available for text analysis. Other

---

[1] http://storage.googleapis.com/books/ngrams/books/datasetsv2.html.

[2] https://tscorpus.com/.

[3] https://ieee-dataport.org/open-access/large-scale-nepali-text-corpus.

[4] https://linguatools.org/tools/corpora/wikipedia-comparable-corpora/.

ways of extracting data are web APIs and web scraping (elaborated in Chap. 2, Sect. 2.6).

3.2.2 Dictionary Creation

It is a sophisticated form of word counting associated with a known lexicon or group of words that carry some type of meaning. Dictionaries are the list of words that correspond to different categories, such as Bing, AFINN, and LIWC, which are popular sentiment dictionaries. You can either:

- **Create your own dictionary**: As an expert, you can develop a meaningful dictionary/lexicon that captures words that you want to see in the text. For instance, if you have millions of tweets and want to search tweets related to economy, then you will prepare a dictionary on the economy subject and add words like economy, unemployment, trade, or tariffs.
- **Use other people's dictionaries**: Such dictionaries are often developed through empirical observation and more rigorous coding techniques, such as Bing dictionary, for sentiment analysis.

When to Use a Dictionary-Based Method?
If you know what you are looking for in a corpus of text, you use a dictionary-based approach, whereas unsupervised methods might be more helpful if you want to do exploratory text analysis.

3.3 Text Pre-Processing

Text pre-processing is one of the critical steps in many text mining tasks that influence the overall mining process to overcome the noise of raw textual data. If pre-processing is not done correctly, it could result in a confounded output making no sense. Careful text pre-processing is needed to ensure the quality of large corpora. Thus, pre-processing of text should be done with great care. Various text pre-processing steps for different text mining approaches and applications are as follows:

3.3.1 Case Normalization

Texts are primarily written in mixed letter cases (uppercase and lowercase). Capitalization helps humans read and distinguish words like nouns from proper nouns and may be helpful in algorithms like entity-named recognition. However, the word's capitalization at the beginning of the sentence is considered identical to the word appearing anywhere in the document with lowercase. For instance, "MANGO," "ManGo," "manGo," and "MAnGo" maps to "mango." Therefore, case normalization helps us to achieve uniformity and standardization of textual documents by converting all the uppercase to lowercase and lowercase to uppercase letters. It helps to solve the sparsity issue and speed up the search process.

3.3.2   Morphological Normalization

This type of normalization converts text into a more convenient and standard form. It helps to improve the quality of text for various text mining and information retrieval processes and is used when there are multiple representations of a single word. The words may mean different but will be similar contextually. For information retrieval, normalization of text includes the process of normalizing the indexed and query terms into the same form. For instance, the word "play," "player," "playing," and "played" can be mapped to "play," where all the disparities of the word are converted into a normalized form. In contrast, for text mining, normalization of text means defining the terms in a generic sense, such as U.S.A is changed to USA. It is an essential step in feature engineering for machine learning, where high-dimensional features are converted to lower-dimensional space.

3.3.3   Tokenization

Tokenization automatically divides a character sequence into smaller units of sentences or words or phrases known as tokens and removes characters like white space or punctuation marks (Fig. 3.3). There are two types:

1. *Word Tokenization*: It is a process of splitting the sentence into words through unique space character but may also tokenize multi-word expressions, such as "New York" to "New" and "York."
2. *Sentence Tokenization*: It is a process of splitting the document into sentences through delimiters like punctuation marks. However, it cannot be implemented

p = "I love watching movies. My favorite movie is Titanic."

Word tokenization : I, love, watching, movies, ., My, favorite, movie, is, Titanic, .

Sentence tokenization : I love watching movies ——— ①
My favorite movie is Titanic ——— ②

Fig. 3.3 Example showing tokenization

to acronyms and abbreviations. For instance, "U.N." would be tokenized into "U" and "N" tokens and lose their meaning but can be overcome by using *smart tokenization* that detects acronyms and abbreviations.

3.3.4　Stemming

Stemming is the process of transforming word tokens to their root/stem form, where suffixes (known as suffix stripping), prefixes, and pluralization such as 'laugh', 'laughs', 'laughed', 'laughing' maps to 'laugh' are removed for derivationally related words with similar meaning. It is language-dependent as different languages have different linguistic rules. It may result in a made-up word, such as "daily" maps to "dai," and, thus, does not always result in base form for its inflectional/variant forms. Stemming algorithms are called *stemmer* that are rule-based, such as Porter (gentle stemmer that removes suffixes), Lancaster (aggressive stemmer), and Snowball (considered as an upgraded version of Porter stemmer) stemmers in the English language. It can lead to the following:

- **Overstemming** results when the word is removed too much and results in nuisance stems where the word's meaning is lost. It can also occur when the different words result in one stem by the stemming algorithm but are from different stems. For instance, "university," "universal," "universities," and "universe" are mapped to the word "univers."
- **Understemming** results when several words are forms of one another but do not map to the same stem. For instance, "data" and "datum" are mapped to "dat" and "datu," respectively.

It is still debatable if stemming enhances the quality of text mining results and efficacy in information retrieval. In text mining, stemming loses contextual information because of the absence of consideration of syntax. In contrast, in information retrieval, for some cases, stemming may increase the efficacy for some queries but may decrease the efficacy for others.

3.3.5　Lemmatization

It is an advanced form of stemming that groups words on their core lemma or concept. Lemma is the base form of all its inflectional or variant forms, that is why standard dictionaries are lists of lemmas and not stems. It performs an arithmetic process to determine the lemma (i.e., dictionary form) by mapping the verb forms to their corresponding infinitive form and nouns to their singular form by understanding the context and parts of speech for a given word in a sentence

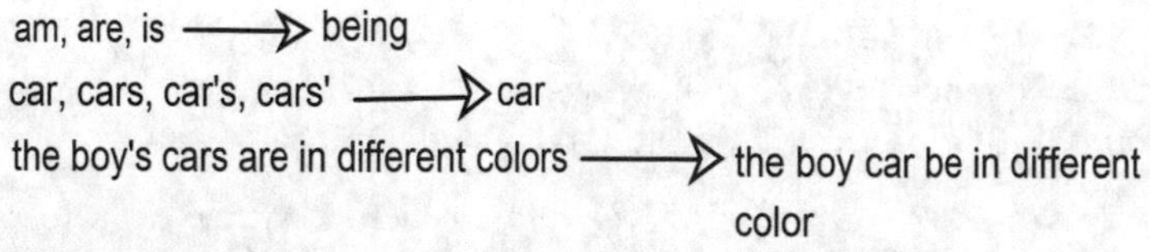

am, are, is ———> being

car, cars, car's, cars' ———> car

the boy's cars are in different colors ———> the boy car be in different color

Fig. 3.4 Example showing lemmatization

(Fig. 3.4). For instance, for the word "meeting," lemmatization will map to the word "meet" for the verb and "meeting" for the noun, whereas in stemming for both verb and noun forms, it will be mapped to the same root "meet." They are of two types: (i) *inflectional morphology*, such as "cutting" maps to "cut" and (ii) *derivational morphology*, such as "destruction" maps to "destroy."

3.3.6 Stopwords

Common words do not carry useful information and occupy extra storage and memory for calculating documents or word vectors. Such words are referred to as *stopwords*, and the procedure to discard them is known as *stopping*. Tokenized words that match with the stopwords from the stopword list are removed. A stopword list is generally preselected and language-dependent. For most text mining tasks and algorithms, removing stopwords is helpful as they have little or no impact on the final results and do not encounter loss of information. In contrast, for text mining tasks, like information retrieval, that involve phrases, removal of phrases may lead to loss of information. In the English language, there are about 500 stopwords.

3.3.7 Object Standardization

The words or phrases that are not in standard lexical dictionary form will not be identified, such as acronyms (dm for direct message, rt for retweet), hashtags with attached words, and slangs. It is generally overcome by using regular expressions[5,6] and manually prepared dictionaries.

---

[5] https://gist.github.com/jacksonfdam/3000275#file-gistfile1-txt.

[6] https://rstudio.com/wp-content/uploads/2016/09/RegExCheatsheet.pdf.

Jordan Patterson is a Cataloguing and Metadata Librarian at the Memorial University Libraries, St. John's, Newfoundland, Canada.

Story: *I learned regular expressions through a Library Carpentry workshop 5 years ago, and now hardly a day goes by when I do not use this tool to extract data from a text file or to reformat or clean data that I have scraped from a website. Since I learned regular expressions in a Library Carpentry workshop, I know that the tool has some degree of uptake in libraries, but considering how useful it is, I do not think nearly enough librarians know about it. Typically I use a website dedicated to regular expressions such as Regexr to explore and manipulate text data, but many programs have built-in support for regular expressions so that you can use them in OpenRefine, Word, and most text editors used for coding, such as Notepad++. For such a wonderful, easy-to-learn tool with broad application and support in many programs, I am surprised I do not hear more about how regular expressions are being used in libraries. I have used regular expressions in cases as simple as reformatting a table of contents for a MARC record and in cases as complex as straightening out the messy data of an entire scraped webpage.*

3.4   Feature Engineering

After the pre-processing of raw text, we can extract features from the processed textual data for further analysis, where each word in a document can be converted to a feature. Feature engineering is the process of encoding text into numeric form to create features so that various text mining and machine learning algorithms can understand the data. Various methods can be used to construct textual features and are summarized below.

3.4.1   Semantic Parsing

It is based on words' syntax, where word order and type are essential. It creates lots of features to study and follows a tree-like structure by breaking the texts into tags (Fig. 3.5).

3.4.2   Bag of Words (BOW)

It treats each term as a single token of a sentence, and the term's type and order are not essential (Fig. 3.6). It uses the frequency of term per document and ignores

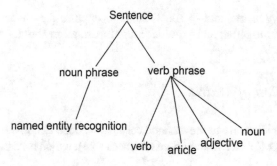

Fig. 3.5 Diagram showing semantic parsing

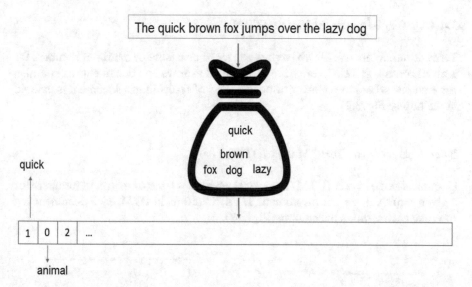

Fig. 3.6 Diagram showing bag of words (BOW)

the term's positions. It significantly simplifies the content of the texts. Many real-world problems and applications use word frequency alone and are sufficient for BOW analysis. Text mining techniques, like Probabilistic Latent Semantic Indexing (PLSA), Latent Semantic Indexing (LSI), and topic modeling, usually use the BOW technique.

3.4.3   N-Grams

N-grams help determine the context where N stands for the consecutive word being used, such as unigram for one term, bigram for two consecutive terms, trigram for three consecutive terms, and so on. It helps to explore the chances where one or more terms occur together and provides more information about the corpus.

3.4.4   Creation of Matrix

To give a proper structure to the cleaned corpus, we need to convert it to document-term matrix (DTM) or term-document matrix (TDM), which uses the raw frequency of the terms.

3.4.4.1   Term-Document Matrix (TDM)

Term-document matrix (TDM) represents terms as a table or matrix of numbers for a given corpus. In TDM, terms are represented as rows and documents as columns for a corpus where the number of occurrences of terms in the document is entered in the boxes (Fig. 3.7).

3.4.4.2   Document-Term Matrix (DTM)

Document-term matrix (DTM) represents terms as a table or matrix of numbers for a given corpus. It is a transposition of TDM; therefore, in DTM, each document is a row, and each word is the column (Fig. 3.8).

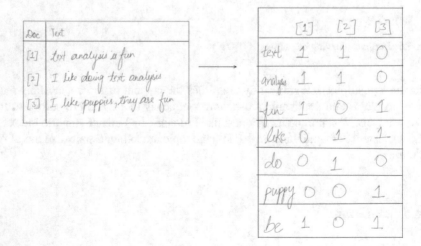

Fig. 3.7 Example showing term-document matrix

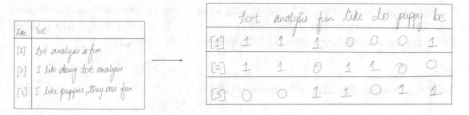

Fig. 3.8 Example showing document-term matrix

3.4.5 Term Frequency-Inverse Document Frequency (TF-IDF)

3.4.5.1 Term Weighting

The term *weighting* is popularly used in information retrieval and supervised machine learning tasks like text classification. It assumes that a word that occurs the most is the word that best describes that particular document. Therefore, it makes a list of more discriminative terms than others and assigns a weight to each highly occurring term.

Why Cannot All Terms Be Considered Equal Based on Their Frequency for a Collection of Documents?
According to Zipf's law, the frequency of the term is inversely proportional to its rank in a corpus, where stopwords are more dominant in a document, and a raw word counting is not very helpful.

3.4.5.2 Term Frequency (TF)

Term frequency (TF) is the frequency of a term (T) in a document (D). It is used in many different applications, such as calculating query-document match scores or creating DTM. Raw or simple term frequency is not a helpful metric as all text corpora follow Zipf's law. For instance, a term occurring 20 times in a document is more relevant than a document consisting of the same term occurring in that document only once, but that does not mean that the term is 20 times more relevant in one document than the other. In text mining and information retrieval, document's relevance when computing *query-document match score* does not increase proportionally with term frequency.

3.4.5.3 Document Frequency (DF)

Rare terms are more discriminative and informative than *frequency terms*. In information retrieval, such terms are weighted more in a document as they are highly likely to be relevant for a query. For a query term, we want to assign positive weights but lower than the rare terms, and this is done by using document frequency (DF). DF is the number of documents (D) that consists of the terms (T).

3.4.5.4   Inverse Document Frequency (IDF)

Inverse document frequency (IDF) estimates the significance of a term in a corpus, where the terms with low DF are more valuable than the terms with high DF:

$$\text{IDF} = \log_{10} \frac{N}{DF} \tag{3.1}$$

Thus, IDF states that the term that occurs more frequently in a corpus is less informative.

3.4.5.5   Term Frequency-Inverse Document Frequency (TF-IDF)

Term frequency-inverse document frequency (TF-IDF) evaluates the relevancy of a term for a document in a corpus and is the most popular weighting scheme in information retrieval. TF-IDF of a term (T) is the product of its TF and IDF weights:

$$\text{TF-IDF} = (1 + \log TF) \times \log_{10} \frac{N}{DF} \tag{3.2}$$

It increases with the number of occurrences and rarity of a term within the corpus, where terms with higher TF-IDF values imply a stronger association with the document they appear in. From an information retrieval perspective, when such terms are used as a query by a user, then the resultant document might be of interest to the user. Thus, it is the de facto method for comparing performance in text mining and information retrieval applications and gives better recall and precision values than other weighted schemes.

3.4.6   Syntactical Parsing

A parse tree (shown inverted with the root at the top) is a graphical representation of the syntactic structure of textual data. It can be constructed using constituency (groups of words behaving as a single unit or constituents) or dependency relation of the grammar. The parse tree consists of morphological structures of the text in a machine-readable format and provides a better structure to analyze text rather than bag-of-words (BOW) representation. It is semantically richer than BOW representation. A sentence can be parsed by relating each word to other words in

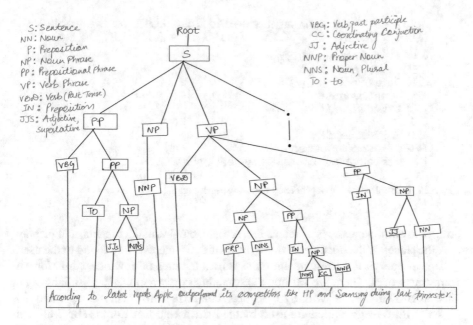

Fig. 3.9 Example showing dependency parsing

the sentence on which they depend. It uses either dependency grammar (DG) or phrase structure grammar (PSG) to find the correct syntactic structure of a given grammar in a sentence. Dependency parsing uses DG more frequently for various natural language processing (NLP) applications than contingency parsing, which uses PSG. Syntactic or dependency parsing represents the grammatical structure of a sentence, which consists of lexical items such as words connected to each other by directed links called dependencies (Fig. 3.9). It is depicted as a tree where nodes represent the words and edges represent the dependencies. It consists of functional categories, which constitute labels and uses some structural categories such as POS.

3.4.7   Parts-of-Speech Tagging (POS)

It is an automatic technique that decides how a word is used and annotates each word in a sentence (Fig. 3.10). It consists of eight main parts of speech—nouns, pronouns, verbs, adverbs, adjectives, conjugations, prepositions, and interjections and other subcategories. It makes dependency parsing effortless and precise as tagging functions much better when graphing and grammar of text are correct. It helps to perform much more effortless and well-organized searches of the corpus. It converts it into a structured database that can be searched using the values of the

John saw the saw and decided to take it to the table
NNP VBD DT NN CC VBD TO VB PRP IN DT NN

NNP: Proper Noun, singular
VBD: Verb, past tense
DT: Determiner
NN: Noun, singular
CC: Coordinating conjuction
TO: *to*
VB: Verb, base form
PRP: Personal pronoun
IN: Preposition or subordinating conjunction

Fig. 3.10 Example showing dependency parts-of-speech tagging

entities to compute indices of relevance, recognize relevant documents, and display specific places in the document where the entities of interest were found to the user.

It is a supervised learning solution as it uses features of previous words for the following words. There are various POS training corpora in the English language, such as Brown Corpus[7] with 87 POS tags, Penn Treebank[8] with 45 tags, and C5 tagset[9] with 61 tags, which are used by the British National Corpus (BNC). Penn Treebank is the most popular POS tagset. Assigning POS tags to words is algorithm-dependent to the corpus that is used to train the algorithm.

POS is most useful in the text to speech, information retrieval, word sense disambiguation,[10] or improving word-based features. POS tagging helps to disambiguate the meaning of the word. For instance, in the sentence "Time flies like an arrow," the word "like" is IN (preposition), but in the sentence "I like candy," the word "like" is VBP (verb). The two methods by which POS can be applied are as follows:

1. **Rule-Based POS Tagging**: The oldest approach that uses hand-coded rules and lexicon/dictionary for tagging. Hand-coded rules are utilized to determine the correct tags when there is a possibility for more than one tag for a word. It is coded in the form of rules based on context-pattern rules or as regular expression.
2. **Learning-Based POS Tagging**: It is more effective than the rule-based method and is trained on human-annotated corpora like Penn Treebank. It can either be based on a statistical model like maximum-entropy Markov model (MEMM), hidden Markov model (HMM), or conditional random field (CRF) or rule-based method, such as transformation-based learning (TBL).

---

[7] http://icame.uib.no/brown/bcm.html.

[8] https://www.ling.upenn.edu/courses/Fall_2003/ling001/penn_treebank_pos.html.

[9] http://www.natcorp.ox.ac.uk/docs/c5spec.html.

[10] Some words have multiple meanings according to their usage in a language, such as the word "go," which can be used as an acronym for gene ontology or a verb.

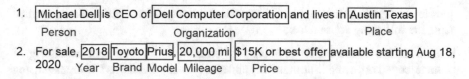

Fig. 3.11 Example showing dependency named entity recognition

3.4.8   Named Entity Recognition (NER)

Named entity recognition (NER) or entity extraction is originated from information extraction but narrower in scope and finds relevant parts in a text, such as names of people, places, or organizations (Fig. 3.11). It collects information from various parts of a text and produces a structural representation of relevant information, for instance, knowledge and relation base. It helps to organize information by providing semantically precise information that permits for additional inferences by algorithms. There can be many entity types that can be used for a problem, but it totally depends on the type of problem for it is used and the NER model being used that provides the entities. If one needs a new type of NER entity that is not present in any model, then they need to develop their own NER training data and train it using the NER algorithm.

NER is a difficult task as:

- The entities are too numerous to be included in the dictionaries.
- Language is changing continuously; similar words might mean different in the present from past.
- The words appear in several different forms.
- Subsequent occurrences of words might be abbreviated, which will not be identified by the NER. For instance, the term "National Science Foundation (NSF)," once defined at the beginning of the document, then the abbreviated form, NEF, will be used in the rest of the document.
- As words can have multiple meanings, this can result in ambiguity.

Various tools can be used for NER, such as Stanford NER,[11] GATE ANNIE,[12] and OpenCalais.[13] Stanford NER model[14] is one of the popular NER models, which consists of a three-class model (location, person, organization), four-class model (location, person, organization, miscellaneous), and seven-class model (location, person, organization, money, percent, date, time). The applications of NER include tasks like information extraction, summarization, natural language understanding,

---

[11] https://nlp.stanford.edu/software/CRF-NER.html.

[12] http://services.gate.ac.uk/annie/.

[13] https://www.drupal.org/project/opencalais.

[14] https://nlp.stanford.edu/software/CRF-NER.html.

question-answering, ontology construction, and many others. NER can be applied using the following approaches:

1. **Knowledge-Based NER**: It needs less amount of training data consisting of hand-coded rules and is domain-dependent. It is very precise as it is created using hand-coded rules but might result in lower accuracy if the rules are not matched for a given text. Further, as it is hand-coded, it is labor-intensive and expensive. As languages evolve, the rules are required to be consequently updated. For instance, *regular expressions* can be created to extract entities like capitalized words, emails, and names from a collection of texts.
2. **Learning-Based NER**: It does not require grammar or rules and gives a higher recall value with a lower precision value. Further, there is no need for linguistic experts to be included as rules or grammar is not used. In comparison to making rules, it is cheaper to make annotations but will need an abundance of high-quality training data as better results are obtained with more training data. It can be performed using the following:

   a. **Supervised Learning Method**: It is a terminal approach that requires labeled training dataset such as hidden Markov model (HMM), K-nearest neighbor, AdaBoost, decision trees, and support-vector machine (SVM).
   b. **Unsupervised Learning Method**: It does not need a training dataset to process, and entities are automatically located without training, such as clustering.

3.4.9   Similarity Computation Using Distances

Text matching is the task that determines the similarity between two documents. It can be performed using the following methods:

1. **Euclidean Distance**: It compares the shortest distance between the documents using the Pythagoras theorem to compute the distance, where 0 indicates identical documents. It is not a good measure for computing distances for multidimensional representations that are sparse and have a significant number of zero values varying significantly over different documents of varying length.
2. **Cosine Distance**: It uses the normalized dot product of each vector representing the documents and does not depend on the length of the vectors. It measures the cosine angle between two vectors, where a smaller angle represents higher cosine similarity. It always lies in the range 0-1, where 1 indicates the same orientation. It computes the geometric mean of the fraction of shared words for a pair of documents.

3. **Jaccard Similarity**: It is another distance metric that uses Boolean representation for the text. For given vectors, it can be computed as the ratio of terms available in both the vectors by the terms available in either of the vectors. It can be represented as:

$$\text{Jaccard Similarity} = \frac{A \cap B}{A \cup B} \tag{3.3}$$

for two vectors, A and B. It ranges from 0 to 1, where 1 indicates documents are identical. Features are created by clustering similar documents, followed by assigning a unique label to each document in a new column.

3.4.10 Word Embedding

It is a technique to identify similarities between words in a text corpus by using a model to predict the co-occurrence of words within a small chunk of text. It redefines word features from high dimensional to low dimensional by preserving the contextual similarity in the corpus. In word embeddings, (i) each word is represented as a vector; (ii) each word is mapped to one vector; (iii) the learning of vector values resembles a neural network. It is considered as a refinement over the bag-of-words model that uses sparse representation. It tries to position the words based on their semantic meaning or context. The most popular algorithms that can be used to create word embeddings are *word2vec*[15] and *GloVe*,[16] whereas *t-SNE* is popularly used to visualize the word embedding.

Word2Vec is a simple "two-layer neural network that is trained to reconstruct linguistic contexts of words" [1] (Fig. 3.12). It can be applied either using a continuous bag of words (CBOW), which looks at a group of words and predicts a specific word, or the skip-gram method, which takes the word and predicts its surroundings. Skip-gram identifies patterns within texts to represent them in multidimensional space, whereas the CBOW is more useful in practical applications, such as predictive web search. Skip-gram performs better than the CBOW method for most text mining tasks, where both methods first construct a vocabulary from the training corpus and then learn word embedding representation from that. It can be used as feature vectors for a machine learning model, text similarity, word clustering, and text classification techniques.

---

[15] https://code.google.com/archive/p/word2vec/.

[16] https://nlp.stanford.edu/projects/glove/.

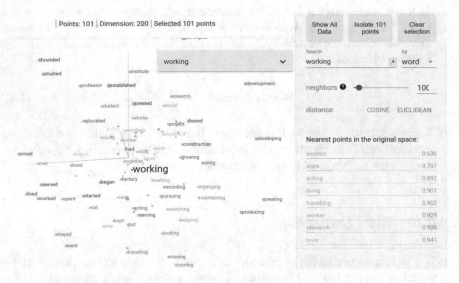

Fig. 3.12 Example showing t-SNE visualization of word2vec using embedding projector (https://projector.tensorflow.org/)

3.5   Case Study: An Analysis of Tolkien's Books

About the Project

Lord of the Rings Project or LOTR Project [2] is a web project run by Emil Johansson, dedicated to J.R.R Tolkien's works. Five books, viz., (a) *The Silmarillion*, (b) *The Hobbit*, and (c) *The Lord of the Rings* trilogy, were used to perform various text analyses. The books were first indexed in a database, followed by cleaning the data by removing indexes and appendices from the book. The LOTR Project [2] website is built using jQuery, Flot, and D3.js.

Word Count and Density

Figure 3.13 presents the bar graphs that compare the frequency of unique words among the books. The *unique word density* graph calculates the percentage of unique words of the whole text, whereas a *unique word* graph shows the comparison of unique words among the books without considering the length of the words. The results showed that "*The Hobbit* book had more unique words than *The Silmarillion* and had the highest unique word density compared to other books. However, this does not necessarily mean that *The Hobbit* book is more difficult to read. *The Silmarillion* and *The Lord of the Rings* both introduce many times more new concepts, characters, and locations, which made them more complex" [2].

Keyword Frequency

Figure 3.14 shows the keyword frequency results for the query "Mordor" for the studied books. The result also includes compound phrases such as Mordorbacon or MordorGandalf in the search. It can be observed from the figure that *The Return of*

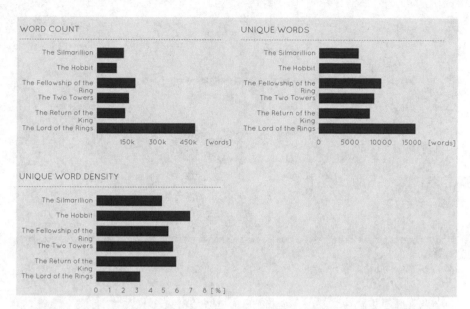

Fig. 3.13 Screenshot showing word count and density graph from the LOTR Project (©LotrProject, all rights reserved—reprinted with permission from Emil Johansson. http://lotrproject.com/statistics/books/wordscount. Accessed 15 July 2020)

the King book had the highest frequency for the keyword "Mondor," followed by *The Two Towers* and *The Fellowship of the Rings*.

Common Words
Figure 3.15 depicts the word cloud for the frequently occurring words in the respective books.

Chapter Lengths
Figure 3.16 shows the graph for the length of an individual chapter for the five books with light gray to black gradient color, where light gray color represented low word count compared to black color, which represented high word count for the chapter.

Word Appearance
"As there are many words, characters, and locations in Tolkien's works, Fig. 3.17 shows where new terms and words appeared on each page for every chapter in the five books. In order to enhance patterns pages where at least 25% of the words are new were shown red color whereas pages with less than 25% new words were shown on a color scale from red to blue" [2].

Sentiment Analysis
"Sentiment analysis is the science of assigning mood to pieces of text based on keywords and structure. A free API called Sentiment140 was used to perform sentiment analysis, which was developed by Alec Go, Richa Bhayani, and Lei Huang" [2]. Figure 3.18 shows the results for sentiment analysis for each sentence, which was then averaged over the page for all the books. In the figure, green color

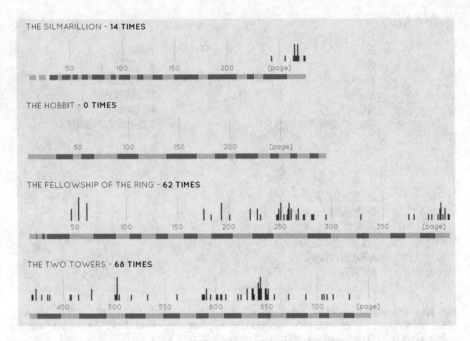

Fig. 3.14 Screenshot showing keyword frequency graph from the LOTR Project (©LotrProject, all rights reserved—reprinted with permission from Emil Johansson. http://lotrproject.com/statistics/books/keywordsearch. Accessed 15 July 2020)

Fig. 3.15 Screenshot showing word cloud of common words from the LOTR Project (©LotrProject, all rights reserved—reprinted with permission from Emil Johansson. http://lotrproject.com/statistics/books/commonwords. Accessed 15 July 2020)

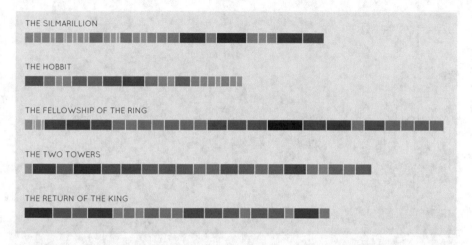

Fig. 3.16 Screenshot showing chapter length of books from the LOTR Project (©LotrProject, all rights reserved—reprinted with permission from Emil Johansson. http://lotrproject.com/statistics/books/chapters. Accessed 15 July 2020)

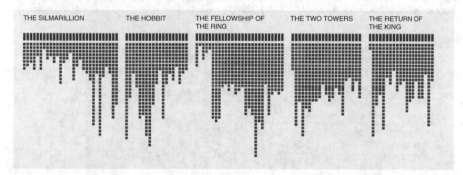

Fig. 3.17 Screenshot showing the appearance of new words from the LOTR Project (©LotrProject, all rights reserved—reprinted with permission from Emil Johansson. http://lotrproject.com/statistics/books/wordappearance. Accessed 15 July 2020)

represents positive sentiment, yellow color represents neutral sentiment, and red color represents a negative sentiment for the respective pages of the books.

Character Co-occurrence
"To visualize the connections between various characters in the books, a force-directed network graph was created for each book. All the volumes for Lord of the Rings books were compiled together to give better insight into the story. The size of the circle represented the total number of mentions for a particular character across the book. In order for minor characters not to disappear completely, both maximum and a minimum possible size of the bubbles were fixed. The lines between two bubbles represent the total number of times both characters were mentioned on the same page (co-occurrence). This method for calculating character interactions

THE SILMARILLION

THE HOBBIT

THE FELLOWSHIP OF THE RING

THE TWO TOWERS

THE RETURN OF THE KING

Fig. 3.18 Screenshot of sentiment analysis of the books from the LOTR Project (©LotrProject, all rights reserved—reprinted with permission from Emil Johansson. http://lotrproject.com/statistics/books/sentimentanalysis. Accessed 15 July 2020)

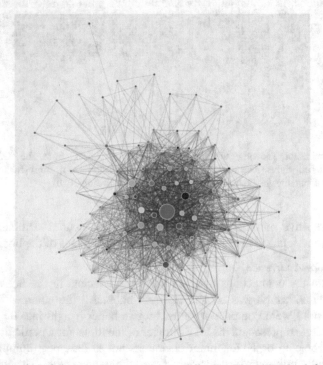

Fig. 3.19 Screenshot of character co-occurrence graph for *The Silmarillion* book from the LOTR Project (©LotrProject, all rights reserved—reprinted with permission from Emil Johansson. http://lotrproject.com/statistics/books/cooccurrences?view=silm. Accessed 15 July 2020)

is not completely optimal since characters are often mentioned on the same page even though they never meet" [2]. Figure 3.19 showed an example of a word co-occurrence graph for *The Silmarillion* book when characters were mentioned on the same page. It was found that Melkor/Morgoth was presented in almost every part of the story with maximum node size.

Appendix B

Language Corpora Available for Text Mining

Table B.1 summarizes selected language corpora available for text mining. Some of the additional datasets which are in open-access and can be used for training and teaching in text mining/NLP are available at https://github.com/textmining-infopros/Curated-Datasets.

Table B.1 Selected language corpora available for text mining

Resource	Description	Fee	Resultformat
Acquis Communautaire (AC) https://ec.europa.eu/jrc/en/ language-technologies/jrc-acquis#Download%20the %20JRC-Acquis%20corpus	It provides access to legislative text between the 1950s to present in 22 languages (Bulgarian, Czech, Danish, German, Greek, English, Spanish, Estonian, Finnish, French, Hungarian, Italian, Lithuanian, Latvian, Maltese, Dutch, Polish, Portuguese, Romanian, Slovak, Slovenian, and Swedish)	Free	Text/XML
Australian National Corpus https://researchdata.edu.au/ australian-national-corpus/2018	It provides access to assorted examples of Australian English text, transcriptions, audio, and audiovisual materials	Free	CSV
BYU Law & Corpus Linguistics https://lawcorpus.byu.edu/	It provides access to the platform for linguistic analysis of the corpus of (i) Founding Era American English, (ii) Early Modern English, (iii) Supreme Court Opinions of the United States, (iv) State Conventions on the Adoption of the Constitution, (v) Records of the Constitutional Convention, (vi) Early Statutes at Large, (vii) US Case Law, and (viii) Current US Code	Free	Google Sheets
Chinese Corpora http://corpus.leeds.ac.uk/query-zh.html	It provides access to Chinese corpora and frequency lists provided by Leeds University	Free	HTML

(continued)

Table B.2 (continued)

Resource	Description	Fee	Resultformat
Chinese-English Parallel Corpora https://www.translatefx.com/resources/corpora	It provides access to aligned sentence pairs from quality bilingual texts, covering the financial and legal domains in Hong Kong	Free	Tab-separated values (TSV)
English-Corpora https://www.english-corpora.org/	It provides access to most widely used online corpora including iWeb, News on the Web, Wikipedia, Coronavirus, Historical American English, The TV Corpus, The Movie Corpus, American Soap Operas, Hansard, Early English Books Online, US Supreme Court Opinions, TIME Magazine, British National, CORE, Google Books n-gram	Paid	Structure Query Language (SQL)
Demo Corpora http://dhresourcesforprojectbuilding.pbworks.com/w/page/69244469/Data%20Collections%20and%20Datasets#corpora	It provides access to toy collections including linguistic corpus that are ready to go for demonstration purposes and hands-on tutorials	Free	Depends on the corpus
European Language Corpora https://www.linguistik.hu-berlin.de/en/institut-en/professuren-en/korpuslinguistik/links-en/korpora_links	It provides access to a variety of corpora in different languages collated by Humboldt University Berlin, Faculty of Language, Literature and Humanities	Free	Depends on the corpus
Japanese Corpora https://www.ninjal.ac.jp/english/database/type/corpora/	It provides access to variety of corpora built by the National Institute for Japanese Language and Linguistics	Free	Depends on the corpus
Parallel Corpora http://www.statmt.org/moses/?n=Moses.LinksToCorpora	It provides access to corpora from different sources	Free	Depends on the corpus
Google Books Ngram Corpus http://storage.googleapis.com/books/ngrams/books/datasetsv2.html	It provides access to n-grams of Google Books	Free	CSV
Wikipedia—list of text corpora https://en.wikipedia.org/wiki/List_of_text_corpora	It provides access to a list of text corpora in various languages collated by Wikipedia	Free	Depends on the corpus

References

1. Qi W, Procter R, Zhang J, Guo W (2019) Mapping consumer sentiment toward wireless services using geospatial Twitter data. IEEE Access 7:113726–113739. https://doi.org/10.1109/ACCESS.2019.2935200
2. Lotr Project: An analysis of Tolkien's books. http://lotrproject.com/statistics/books/. Accessed 15 July 2020

Chapter 4
Topic Modeling

Abstract Topic modeling is usually used to identify the hidden theme/concept using an algorithm based on high word frequency among the documents. It can be used to process any textual data commonly present in libraries to make sense of the data. Latent Dirichlet Allocation algorithm is the most famous topic modeling algorithm that finds out the highly contributing words in each topic. Additionally, it provides a topic proportion that can segregate all the documents under the identified themes/topics. Thus, topic modeling can tag each document with a topic that can later be used to index and link the defined set of documents if embedded in the website or database for better searching and retrieval purpose. A comprehensive conceptual framework related to topic modeling, tools, and ways to visualize topic models is covered in this chapter with various use cases. This chapter is followed by a case study using three different tools to demonstrate the application of topic modeling in libraries.

4.1 What Is Topic Modeling?

Topic modeling makes an excellent tool for discovery and helps to uncover evidence already present in the text. A *topic* can be defined as the main idea discussed in a text, i.e., the theme or subject of different granularities. In contrast, topic modeling acts as a text mining approach to understand, organize, process, extract, manage, and summarize knowledge. It has been called an act of reading tea leaves[1] or the process of highlighting words[2] based on their topics. It is based on statistical and machine learning techniques to mine meaningful information from a vast corpus of unstructured data and is used to mine document's content. There are no machine-readable annotations that can tell the topic modeling programs about the semantic meaning of the words in the text. Thus, it infers abstract topics based on "similar

---

[1] https://papers.nips.cc/paper/3700-reading-tea-leaves-how-humans-interpret-topic-models.pdf.

[2] https://journalofdigitalhumanities.org/2-1/topic-modeling-a-basic-introduction-by-megan-r-brett/.

patterns of word usage in each document" [1]. These topics are simply groups of words from the collection of documents that represents the information in the collection in the best way [2]. Topic modeling helps to detect patterns like word frequency or distance between words. It clusters documents based on their similarity and the words that occur frequently. It results in the clustering of documents based on their similarity and the clustering of words that infer that relationship. It performs soft clustering, where it presumes that every document is composed of a mixture of topics. It attempts to identify the topics that are presented in the documents and their presence.

Topic modeling is a model of texts that is built with a particular theory in mind but cannot provide evidence for that theory, whereas typical theories are built into the model's assumptions with a hope that the model will help to point to such evidence [3, 4]. There are mainly two tasks performed in topic modeling—first to discover the major topics in text data and second to analyze which documents cover which topics. Thus, topic modeling helps (i) to organize, search, summarize, and understand extensive collections of text documents, (ii) to identify latent topical patterns present in the collection(s), and (iii) to annotate the documents with the modeled topics. The process of topic modeling includes data collecting, data pre-processing, and model training. The common steps to perform topic modeling are as follows:

- Preparing a corpus (such as converting files from PDF to plain text format)
- Conducting text pre-processing (covered in Sect. 3.3)
- Determining the number of topics (using perplexity, coherence, entropy)
- Selecting the appropriate algorithm (such as latent Dirichlet algorithm, structural topic modeling, or correlated topic modeling) and learning method (unsupervised, supervised, or semi-supervised)
- Seeding (so that one can reproduce the algorithm with the same selected parameters)
- Running the selected algorithm using proprietary or open-source tools (such as RapidMiner, TopicModelingTool) or programming languages (such as R or Python)
- Iterating the whole process till the algorithm fits the model

4.1.1   Topic Evolution

The concept/theme of a text corpus evolves with time. "As time passes, topics in a document corpus evolve, modeling topics without considering time will confound topic discovery. Modeling topics by considering time is called topic evolution modeling. Topic evolution modeling can disclose important hidden information in the document corpus, allowing identifying topics with the appearance of time, and checking their evolution with time" [5]. For instance, one can map the topic lineage in scientific literature and determine its influence on a topic by checking its topic evolution. Dynamic topic models (DTM), topic over time (TOT), dynamic topic correlation detection, multiscale topic tomography, and structural topic model are some of the algorithms that perform topic evolution and consider *time* an essential factor. Lamba and Madhusudhan [1] describe that topic evolution assists in identifying "topics within a context and how they advance in time. Over time, few documents within a topic may initiate content that varies from the original content. If that initiated content is shared by a lot of later documents, the content is recognized as a new topic. Hence, with the progression of time, topics advance, new themes emerge, and old ones become obsolete." They also emphasized that "topic modeling not just helps the researchers to decide the trending topics or related fields to their field of intrigue but additionally encourages them to distinguish new concepts and fields over time" [1].

4.1.2 Application and Visualization

"Besides classical text document analysis and genetics, topic modeling has turned out to be of use in bioinformatics, digital libraries, recommender systems, computing in the service of political and social studies (*digital humanities*)" [6], computer science, medicine, and other application areas. It can easily be generalized to different kinds of data, such as biological data, images, musical notes,[3] survey data, and information. Although topic modeling provides valuable insight in surfacing the latent semantic structures across collections, it should be used to tell a story but not to predict language. If evaluated to determine its ability to tell a story to others, then it should be measured directly by measuring how interpretable the model is and how much a human can make sense of the topic model.

It has a wide range of applications, such as tag recommendation,[4] text categorization, keyword extraction, similarity search, exploratory analysis, discovery, browsing, text summarization, marketing,[5,6] literature review,[7] citation analysis,[8,9] improving search results,[10] and organizing documents. "Some of the important open questions in topic modeling have to do with: How we use the output of the algorithm? How should we visualize and navigate the topical structure? What do the topics and document representations tell us about the texts?" [3].

Places such as libraries, where "texts are paramount, are an ideal testbed for topic modeling and fertile ground for interdisciplinary collaborations with computer

---

[3] https://cseweb.ucsd.edu/~dhu/docs/nips09_abstract.pdf.

[4] https://link.springer.com/article/10.1007/s11192-019-03137-5.

[5] https://link.springer.com/article/10.1007/s11573-018-0915-7,

[6] https://doi.org/10.1177/1470785319863619.

[7] https://doi.org/10.1186/s40537-019-0255-7.

[8] https://doi.org/10.1002/asi.22883.

[9] https://doi.org/10.1145/1645953.1646076.

[10] https://www.bad.pt/publicacoes/index.php/cadernos/article/view/2034/pdf#.

scientists and statisticians. The output of topic modeling is not entirely human-readable, and one way to understand the results is through visualization (see Chaps. 9 and 10) but be sure that you can understand the visualization as topic modeling tools are fallible" [3], and whether the parameters, method, or algorithm chosen by you is not right, then it may return some atypical results. "Topic models are meant to help interpret and understand texts, but it is still the researcher's job to do the actual interpreting and understanding" [3].

Many papers [7–14] studied the different ways of visualizing the results from topic modeling. Topic visualization is a subfield in itself and has been thoroughly studied. Some of the websites/dashboards that show the working examples of topic modeling visualizations are The National Academic Press,[11] iArxiv,[12] InPhO Topic Explorer,[13] TopTom,[14] POLO,[15] Shodhganga ETDs,[16] AIIMS Delhi COVID-19 Research,[17]   COVID-19 Research-CORD -19 Dataset,[18]   Visualizing COVID-19 Research Literature (Tableau),[19] Research.Scot,[20] 9 Years of UK \& EU Research,[21] and Stanford Dissertation Browser.[22]

4.1.3   Available Tools and Packages

There are many open-source tools/applications, such as Topic Modeling Tool,[23] Topix,[24] RapidMiner,[25] VyontTools,[26] WEKA,[27] ConText,[28]

---

[11] https://www.nap.edu/academy-scope/#top-downloads.

[12] https://iarxiv.org/.

[13] https://www.hypershelf.org/.

[14] https://labs.densitydesign.org/toptom.

[15] https://github.com/ontoligent/polo.

[16] https://public.tableau.com/app/profile/manika6395/viz/Shodhganga/Dashboard.

[17] https://manika-lamba.github.io/stm/.

[18] https://strategicfutures.org/TopicMaps/COVID-19/cord19.html.

[19] https://engineering.tableau.com/visually-exploring-the-covid-19-research-literature-6ff2f70035cb.

[20] https://strategicfutures.org/TopicMaps/.

[21] https://strategicfutures.org/hexmaps/ukeutopics/2017-01/app/.

[22] https://nlp.stanford.edu/projects/dissertations/info.html.

[23] https://code.google.com/archive/p/topic-modeling-tool/.

[24] https://topix.io/explore.html.

[25] https://rapidminer.com/.

[26] https://voyant-tools.org/.

[27] https://www.cs.waikato.ac.nz/ml/weka/.

[28] https://context.lis.illinois.edu/publications.php.

DARIAH Topics Explorer,[29] ORANGE,[30] and jsLDA,[31] that can be utilized by non-programmers to perform topic modeling. Currently, there are several programming languages and environments available for programmers to conduct topic modeling. Some of the popular packages in R to perform topic modeling include quanteda,[32] stm,[33] tm,[34] lda,[35] topicmodels,[36] text2vec,[37] topicdoc,[38] BTM,[39] tidytext,[40] and textmineR,[41] whereas some of the popular packages in Python for topic modeling include scikit-learn,[42] corextopic,[43] genism,[44] lda,[45] tethne,[46] and dynamic_topic_modeling,[47] Most of the tools mentioned above are covered in detail in Chap. 10. Programmers can refer to the website[48] and GitHub account[49] of David M. Blei lab for additional topic modeling software.

4.1.4 When to Use Topic Modeling

- When you have a vast collection of text documents
- When the collection belongs to a specific subject
- When the collection has a similar type of documents, such as when all files in the collection are newspaper articles

---

[29] https://dariah-de.github.io/TopicsExplorer/.

[30] https://orangedatamining.com/.

[31] https://mimno.infosci.cornell.edu/jsLDA/.

[32] https://quanteda.io/.

[33] https://cran.r-project.org/web/packages/stm/vignettes/stmVignette.pdf.

[34] https://cran.r-project.org/web/packages/tm/tm.pdf.

[35] https://cran.r-project.org/web/packages/lda/lda.pdf.

[36] https://cran.r-project.org/package=topicmodels.

[37] https://cran.r-project.org/web/packages/text2vec/index.html.

[38] https://cran.r-project.org/web/packages/topicdoc/index.html.

[39] https://cran.r-project.org/web/packages/BTM/index.html.

[40] https://cran.r-project.org/web/packages/tidytext/vignettes/tidytext.html.

[41] https://cran.r-project.org/package=textmineR.

[42] https://pypi.org/project/scikit-learn/.

[43] https://pypi.org/project/corextopic/.

[44] https://pypi.org/project/gensim/.

[45] https://pypi.org/project/lda/.

[46] https://pythonhosted.org/tethne/.

[47] https://pypi.org/project/dynamic-topic-modeling/.

[48] https://www.cs.columbia.edu/~blei/topicmodeling_software.html.

[49] https://github.com/Blei-Lab.

4.1.5   *When* **Not** *to Use Topic Modeling*

- When you have a relatively small number of documents.
- When you do not have any idea about your collection. In this case, clustering will be a better option than using topic modeling.
- When the collection has a mixture of different types of documents, such as when the collection is composed of newspaper archives, journal articles, and ETDs.

4.2   Methods and Algorithms

"Topic modeling is a probabilistic generative model that has been used widely in the field of computer science with a specific focus on text mining and information retrieval in recent years. Latent Semantic Indexing (LSI) served as the basis for the development of topic modeling. Nevertheless, LSI is not a probabilistic model and not an authentic topic model. Based on LSI, Probabilistic Latent Semantic Analysis (PLSA) was proposed" [15]. In 2003, Blei, Ng, and Jordan were the first to describe the widely used topic modeling technique called Latent Dirichlet Allocation (LDA) algorithm [16]. "LDA is an even more complete probabilistic generative model and is the extension of PLSA. Nowadays, a growing number of probabilistic models are based on LDA via combination with particular tasks" [15].

LDA specifies a generative process of imaginary probabilistic *recipe* (Figs. 4.1 and 4.2a). It uses several documents and presumes that the words are related in each document to determine the underlying topics in text documents presented as patterns of tightly co-occurring terms, where the frequently occurring words tend to be about the same subject. In other words, it tries to understand the *recipe* for how each document could have been created. However, in LDA, we are also required to specify the number of topics to build the model, which is then used by the *recipe* to create the word distributions and topics for the corpus. Finally, based on the output, we can determine similar documents within the corpus. The assumptions of LDA are "(i) documents with similar topics will use similar groups of words, (ii) documents are a probability distribution over latent topics, and (iii) topics are probability distributions over words" [17].

"A topic represents the broader concept shared by a document corpus" [1]. The *topic representations (or topic proportions)* of the documents can be used in many ways to analyze the collection. They can be used to (i) "isolate a subset of texts based on the topic of interest, (ii) examine the words of texts themselves and restrict the attention to particular words of interests, (iii) find similarities between them or trends" [3], (iv) determine the frequency of topics, (v) cluster the topics, (vi) determine the correlation between the topics to identify the relationship of topics to each other in practice, and (vii) visualize the topics using heatmaps, topic networks, and other types of visualizations. We need to know the topics and the composition of topics for those documents to perform the above analyses. LDA ignores syntactic

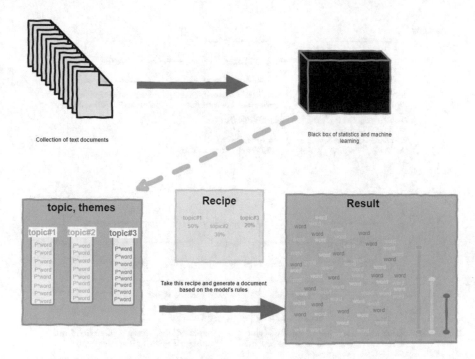

Fig. 4.1 Diagram showing the process of topic modeling

information of the documents and the ordering of the words. It aims to identify the mixture of topics that it comprised of by treating them as *bags of words* (BOW). BOW is good at capturing the subject matter of documents but not nuance. LDA outputs the text as a table of tokens (Fig. 4.2b (i)) that helps to generate two other tables (Fig. 4.2b (ii)): one showing the number of topics used in each document (*DOCTOPIC—θ*) and other showing the number of words associated with each topic (*TOPICWORD—φ*). These tables are then converted into a set of probability distributions (Fig. 4.2b (iii)). Figure 4.2b (iii) shows the average topic weights for documents labeled by authors in a topic model generated from a corpus.

The advantages of LDA include the fact that it is not only a powerful tool for topic modeling but also the easiest one among other topic modeling algorithms and also the easiest to understand conceptually. It has been shown to give good results over numerous knowledge domains and has been used as a vital building block to provide new applications or modify the old ones. The limitations of LDA "include the (i) prior identification of an appropriate number of topics for the documents before performing the algorithm; (ii) incompetence of the Dirichlet topic distribution to correlate among topics; and (iii) manual interpretation and labeling of topics. Although some topics are fairly straightforward to label, others prove more difficult to ascertain the content or relationship that connect the words" [1].

Hence, many algorithms can be used to perform topic modeling, including Latent Dirichlet Allocation (LDA), probabilistic latent semantic analysis (PLSA),

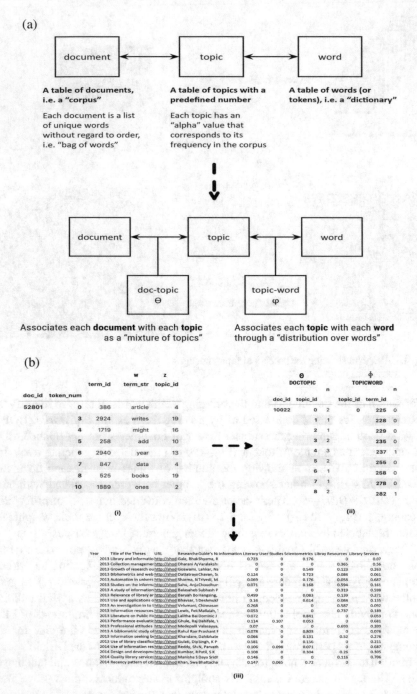

Fig. 4.2 Representation of text in LDA. (**a**) Relationship of main entities of topic model. (**b**) Representation of the entities as table

latent semantic analysis (LSA), correlated topic modeling (CTM), hierarchical topic modeling, structural topic modeling (STM), and correlation explanation (CorEx). This chapter elaborated more on LDA as it is the most popular algorithm and is readily available for use by people with and without programming knowledge. Further, topic modeling can be conducted in an unsupervised, semi-supervised, or supervised learning manner. The LDA method that is explained so far is an unsupervised method as it is applied to a collection of documents without predefined list of words/tags. In contrast, if we guide the LDA with a dictionary or seed words per topic to cover the topics in a particular direction, then it will be a supervised method of performing LDA (such as `guidedLDA`[50]). In contrast, the semi-supervised learning method is somewhere between unsupervised and supervised methods. In semi-supervised method, we first run the LDA in an unsupervised manner and look for some trend or pattern and then guide the model with some anchor or seed words based on the observed pattern.

4.3 Topic Modeling and Libraries

As a librarian, you want to (i) organize your website, institutional repository, in-house database, emails, or reference desk questions or (ii) know what your patrons are saying about your resources, workshops, book clubs, or particular features of a new service you launched, recently. Then rather than spending hours searching stacks of posts, feedback, and chats, you can analyze the desired textual data using topic modeling tools in an effort to identify the texts that cover particular topics. Thus, if you want to analyze texts and identify texts' topics, then you should use topic modeling.

Topic modeling has been applied to numerous resources, such as annual meetings,[51] diary,[52] clinical notes,[53] case reports,[54] newspapers,[55,56,57,58] journals,[59,60]

---

[50] https://github.com/vi3k6i5/guidedlda.

[51] https://doi.org/10.3141/2552-07.

[52] https://www.cameronblevins.org/posts/topic-modeling-martha-ballards-diary/.

[53] https://www.biorxiv.org/content/10.1101/062307v1.full.pdf.

[54] https://doi.org/10.1016/j.compbiomed.2019.04.008.

[55] https://doi.org/10.1002/asi.20342.

[56] https://doi.org/10.1007/978-3-030-39098-3_9.

[57] https://dsl.richmond.edu/dispatch/Topics.

[58] https://www.columbia.edu/~jwp2128/Papers/HoffmanBleiWangPaisley2013.pdf.

[59] https://doi.org/10.1111/bjet.12907.

[60] https://www.ncbi.nlm.nih.gov/pmc/articles/PMC6722707/.

research articles,[61] preprints,[62] patents,[63,64,65,66] conferences,[67] chats, online reviews[68,69] MOOCs,[70] call for papers,[71] social media platforms,[72,73,74,75,76] RSS feed,[77] blogs,[78,79] open-ended survey responses,[80,81] and emails.[82,83] In libraries, topic modeling can be applied to digital libraries' resources;[84] smart card data;[85] EZproxy daily log files,[86] data from library mobile apps,[87] virtual libraries' resources, chats, and reference questions; library databases; in-house journals; institutional and digital repository resources; theses and dissertations[88,89,90] and WebOPACs; MOOC feedback, chats, and suggestions; online library chats and forums;[91] emails; syllabuses; and library's social media platform accounts.[92]

---

[61] https://doi.org/10.1186/s13673-019-0192-7.

[62] https://doi.org/10.1002/asi.23347.

[63] https://doi.org/10.1080/09537325.2019.1648789.

[64] Xu, S., Zhu, L., Qiao, X., Shi, Q., & Gui, J. (2012). Topic linkages between papers and patents. Proceedings of the 4th International Conference on Advanced Science and Technology, 176–183.

[65] https://doi.org/10.1007/s11192-014-1328-1.

[66] https://doi.org/10.1002/asi.24175.

[67] https://arxiv.org/ftp/arxiv/papers/1912/1912.13349.pdf.

[68] https://doi.org/10.1177/1470785319863619.

[69] https://www.mdpi.com/2071-1050/12/8/3402/htm.

[70] https://doi.org/10.1007/978-3-030-02925-8_29.

[71] https://doi.org/10.1108/LHT-02-2019-0048.

[72] https://arxiv.org/abs/2004.02566v3.

[73] https://doi.org/10.1177/0165551515608733.

[74] https://doi.org/10.1007/s10489-019-01438-z.

[75] https://ceur-ws.org/Vol-1391/119-CR.pdf.

[76] https://www.cs.jhu.edu/~mdredze/publications/aaai16_collective.pdf.

[77] https://digitalcommons.mtu.edu/etds/801/.

[78] https://blog.codecentric.de/en/2017/01/topic-modeling-codecentric-blog-articles/.

[79] https://doi.org/10.1016/j.eswa.2010.10.025.

[80] https://economics.expertjournals.com/23597704-605/.

[81] https://doi.org/10.1109/ACCESS.2020.2974983.

[82] https://blog.echen.me/2011/06/27/topic-modeling-the-sarah-palin-emails/.

[83] https://static.googleusercontent.com/media/research.google.com/en//pubs/archive/34948.pdf.

[84] https://doi.org/10.1145/1816123.1816156.

[85] https://doi.org/10.1016/j.trc.2020.102627.

[86] https://osf.io/utgev/.

[87] https://doi.org/10.1108/LHT-05-2018-0066.

[88] https://www.bad.pt/publicacoes/index.php/cadernos/article/view/2034/pdf.

[89] https://doi.org/10.5281/zenodo.3545907.

[90] https://doi.org/10.5281/zenodo.1475795.

[91] https://doi.org/10.4018/978-1-5225-9373-7.ch006.

[92] https://doi.org/10.1002/pra2.2018.14505501024.

Karen Harker and **Sephra Byrne** are University Librarians at the University of North Texas, Texas, United States. They have applied information visualization and clustering techniques in their library.

Story: Subject Evaluation for Collection Mappings
At the University of North Texas Libraries, subject-based collections are evaluated on a rotating basis in order to ensure that the diverse needs of each academic department are being met. We start each review by researching the current program to understand the needs, particularly regarding subjects covered in curricula and research.

Manually analyzing the content of courses, dissertations, and research would not only be labor-intensive but also be susceptible to both errors and bias. To improve both the efficiency of the process and the quality, we have started this year using text clustering to identify key subjects. Our processes are evolving, but they follow standard text clustering techniques of inputting source data, setting and adjusting clustering parameters, and then visualizing results.

First, each course description is collected from the university catalog into a spreadsheet. The data is then cleaned by removing irrelevant text (e.g., course number or "3 s.h."). We then use a Python text mining program to perform a process called tokenization that turns each continuous string into individual tokens for each term. These tokens are filtered to remove words that could skew the analysis like conjunctions and articles, the characters are put in lower case, and then they are stemmed (e.g., "learned" and "learning" are stemmed to "learn").

Once the data set has been cleaned and tokenized, it is clustered. This process uses term frequency-inverse document frequency (TF-IDF) to assign a numerical weight to each term based on its rarity and, therefore, importance in a document and then uses k-means clustering to sort those weighted terms into clusters. There is much in the literature that describes this method in more detail than we have space for in this paper. We export the data set and then import it into a visualization software program like Tableau to create word clouds for each cluster. Each word cloud is then reviewed to determine the overarching theme of each cloud. We often need to run this several times with different numbers of clusters to find the balance of clusters that capture all relevant subjects without being overwhelming. After this is finished, these clusters are compared against the current subjects assigned to collection to determine gaps and irrelevant subjects.

Initially, we used simple word clouds and then migrated to using simple text mining using RapidMiner, and we now create clusters with Python, but we still have room to improve. For example, we would like to automate the initial data collection to save time, as well as assign different weights to different courses

(continued)

to account for class size and level. Overall, the hope is to use this project to improve our collection evaluation process and continue to empower our patrons with the resources that they need for success.

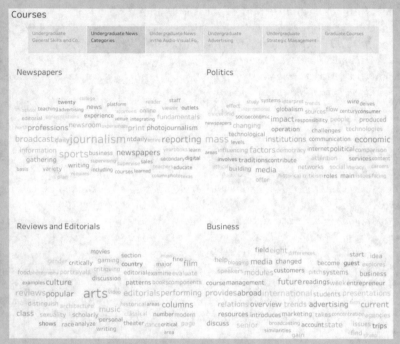

These word clouds are an example from a recent subject evaluation for Journalism and Strategic Communications. The full viz can be found at https://public.tableau.com/profile/untlibraries#/vizhome/Journalism_Colle ctionEvaluation2020/Courses

Rachel Miles is a University Librarian at Virginia Polytechnic Institute and State University, Virginia, and caters the needs of humanities, social sciences, STEM, fine arts, and professional discipline users.

Story: *The work I do focuses on visualizing scholarly publication and citation data with VOSviewer, a software tool for visualizing bibliometric networks via exported files from bibliographic databases, such as Web of Science and Scopus. My story focuses mainly on using the tool to create term co-occurrence maps, in which VOSviewer identifies text data in the publication titles and abstracts using natural language processing algorithms. A network*

(continued)

> *of terms (represented by nodes) represents the number of documents in which a term occurs, while the links between the items represent the number of documents in which two terms occur together. Nodes and lines are color-coded, and depending on the analysis, the colors can represent citation impact, year of publication, or relatedness of research. These visualizations offer a window into the construction and network of scholarly communication for a field of research or a group of researchers, such as a college or department. These visuals can offer insight into a department's research focuses and areas of impact, for instance.*

4.3.1   Use Cases

4.3.1.1   Making Ontologies

Ontologies are one of the most crucial library tasks that have been used in the generation of grammar, query expansion, information extraction, and retrieval, but building a comprehensive ontology is very challenging and time-consuming. Such semantic resource takes a considerable amount of human effort and is a significant drawback from developers' perspective. Topic modeling can act as an efficient method to prepare ontologies by reducing the efforts and time to develop them. A study by Mehler and Walitinger [18] used topic modeling to build a Dewey Decimal Classification (DDC)-based topic classification model in digital libraries. They explored "metadata provided by the Open Archives Initiative (OAI) to derive document snippets as minimal document representations.

Further, they performed feature selection and extension using social ontologies and related web-based lexical resources. This was done to provide reliable topic-related classifications while circumventing the problem of data sparseness" [18]. Finally, they "evaluated the model utilizing two language-specific corpora. The paper bridged digital libraries, on the one hand, and computational linguistics, on the other. The aim was to make accessible computational linguistic methods to provide thematic classifications in digital libraries based on closed topic models such as the DDC. The approach took the form of text classification, text-technology, computational linguistics, computational semantics, and social semantics" [18].

4.3.1.2   Automatic Subject Classification

As topic modeling identifies core topics in a corpus of documents, thus, they can be used in libraries to index subject terms for the studied documents. Studies such as [19, 20] used topic modeling to index the documents.

4.3.1.3   Recommendation Service

Topic modeling can be used to recommend electronic resources (using topic proportion methods) based on the reading or search habits of the users. Lamba and Madhusudhan conducted few studies [1, 17, 21, 22] where they tagged the electronic resources (theses, dissertations, or journal) published in LIS by using topic modeling and determined the core topics for the corpus. All the resources were then annotated using topic proportion. Based on the reading or search habits of the users, they suggested that recommendation service can be provided to users. It can be done by embedding the topic proportion results into the digital library website or the website of the publisher/aggregator or by directly sending the list of recommended resources to patrons by SMS or email based on their weekly/monthly reading habits.

4.3.1.4   Bibliometrics

Topic modeling is an instrumental methodology in the domain of bibliometrics[93],[94],[95] to study evolutionary pathways,[96],[97] citations,[98] and trends[99] to explore different hot and cold topics of research in a particular discipline.

4.3.1.5   Altmetrics

Topic modeling can help the libraries to determine the concepts or keywords appearing in the altmetric data to (i) find the public acceptance of your university research such as vaccine compared to another competitor or peer working in the industry, (ii) determine how people are talking about your university/institute, (iii) track engagement with the services and products provided by the libraries and analyze the topics surrounding the research areas of your university, (iv) predict the future impact applications of your university research based upon what public is saying online, (v) understand your library/university audience who are tweeting/posting about your library/university and use the context of the tweets to cluster the audiences into highly specific user segments and also how they are engaging with the university research and which topics of research are trending [23], (vi) develop communication plans for promoting research on social media, and (vii) outreach to library advocacy groups. You can also answer questions like how people

---

[93] https://doi.org/10.1027/2151-2604/a000318.

[94] https://doi.org/10.1007/s11192-014-1321-8.

[95] https://escholarship.org/uc/item/0zm8h881.

[96] https://doi.org/10.1002/asi.23814.

[97] https://doi.org/10.1103/PhysRevPhysEducRes.16.010142.

[98] https://clgiles.ist.psu.edu/pubs/CIKM2009-topic-evolution-citations.pdf.

[99] https://dl.acm.org/doi/10.5555/1613715.1613763.

talking about research on Twitter is different from mainstream media or bloggers, what topics they care about, and do people care about different topics based on their location.

4.3.1.6   Organization and Management of Resources

The meta-tagging of the electronic resources using topic modeling and topic proportion "not only saves the time of the users but also helps in organizing and management" [1] of the electronic resources. It is the most efficient way to organize and manage a library's database, website, and repository resources. A topic modeling tags a resource based on the theme/concept behind it; it can be a very useful TDM technique for librarians to increase the visibility and usability of their library resources. Thus, topic modeling also helps in knowledge organization and management.[100]

4.3.1.7   Better Searching and Information Retrieval of Resources

In digital libraries, topic modeling also helps in providing a fast searching experience to users and better information retrieval of electronic resources.[101]

4.4   Case Study: Topic Modeling of Documents Using Three Different Tools

4A: Topic Modeling Tool

About the Case Study This case study has been adapted from Lamba and Madhusudhan [1] to illustrate a topic modeling problem using topic modeling tool (TMT) and provides a step-by-step tutorial to solve it.

Problem "Scientific research is highly dynamic in nature, and it is hard to keep an eye on the dynamic development of one's own field, much less all domains under a discipline. The topics explored from research articles will have a direct effect on the development of a field" [1]. Hence, this case study analyzed full-text research articles from an Indian open-access journal for the period of 38 years using Latent Dirichlet Allocation.

---

[100] https://doi.org/10.5771/0943-7444-2018-2-170.

[101] https://www.aclweb.org/anthology/C10-2134.pdf.

Fig. 4.3 Screenshot showing
Steps 1 and 2

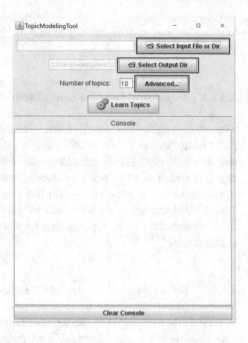

Dataset A folder containing plain text files[102] for 928 research articles retrieved from DESIDOC Journal of Library and Information Technology (DJLIT) during 1981–2018.

About the Tool Refer to Chap. 10, Sect. 10.2.2, to know more about Topic-Modeling-Tool (TMT).

Theory MAchine Learning for LanguagE Toolkit (MALLET)[103] is a Java-based statistical package that analyzes large collections of unlabeled text. It uses an implementation of Gibbs sampling to optimize document-topic hyperparameter and helps to import, train, and infer topics in TMT.

Methodology The following screenshots demonstrate the steps which were taken to perform topic modeling:

Step 1: The folder with particular time-slice data was loaded, followed by selecting an empty output folder where the generated files were saved (Fig. 4.3).
Step 2: Number of topics (K) were tweaked to fit the data (Fig. 4.3).
Step 3: The number of words was then tweaked to fit the data in the new dialogue box that was opened after clicking the "Advanced" option (Fig. 4.4).

---

[102] https://github.com/textmining-infopros/chapter4/tree/master/4a_dataset/.

[103] https://mallet.cs.umass.edu/topics.php.

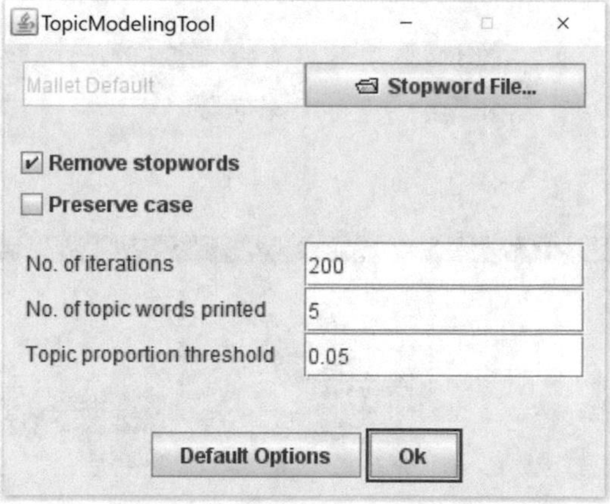

Fig. 4.4 Screenshot showing
Step 3

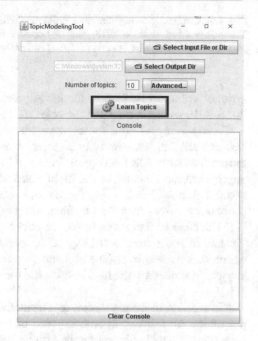

Fig. 4.5 Screenshot showing
Step 4

Step 4: Finally, the Mallet topic modeling algorithm was run by clicking the "Learn Topics" button. The output files were automatically saved in the selected output directory (Fig. 4.5).

Results Eight time slices were finalized based on the volume distribution of DJLIT articles. "For each time-slice, a different number of topics were chosen to fit the distribution of research articles. Accordingly, the periods 1981–1990; 1991–1995; and 2000–2003 were not modeled from Topic-f onwards while the periods 1996–1999; and 2004–2007 were not modeled from Topic-g onwards based on the

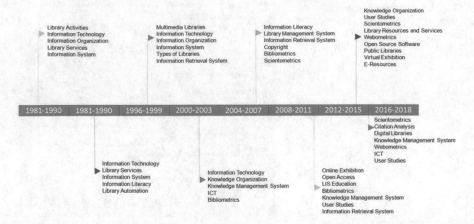

Fig. 4.6 Timeline showing the core topics in DESIDOC Journal of Library and Information Technology from 1981 to 2018 (©2019 Springer Nature, all rights reserved—reprinted with permission from Springer Nature, published in Lamba and Madhusudhan [1])

articles' volume distribution for those years. Also, the periods 2008–2011; and 2016–2018 were not modeled for `Topic-h` and `Topic-i`" [1].

Furthermore, 50 core topics were identified that fitted the corpus of DJLIT research articles, wherein only 29 topics were identified as unique (Fig. 4.6). The output files from TMT recognized the top five words for each topic and assigned appropriate topic names to the highly ranked words. Table 4.1 sums up the LDA result and shows the labeling for the topics, a through i (where a had the highest probability value) in a "descending order according to their probability values" [1]. The table further summarizes the core topics that were the hot research trends "in LIS in India for the corresponding period. The table also lists the word co-occurrence pattern over time and sums up the top five words or the high loading keywords, ranked by the likelihood value for every period in the descending order" [1].

"In addition to the top words, representative articles were also consulted to label the topics. An article can be composed of a single or a mixture of topics, but the core topic is decided based on the articles with the highest value of the topic" [1] proportion for the modeled topic. The supplementary file[104] presents the percentage composition for each studied article.

---

[104] https://github.com/textmining-infopros/chapter4/blob/master/4a_supplementary.docx.

Table 4.1 Extended Latent Dirichlet Allocation topic and word result (©2019 Springer Nature, all rights reserved—reprinted with permission from Springer Nature, published in Lamba and Madhusudhan [1])

Labels	1981–1990 (N = 48)	1991–1995 (N = 46)	1996–1999 (N = 94)	2000–2003 (N = 55)	2004–2007 (N = 79)	2008–2011 (N = 202)	2012–2015 (N = 238)	2016–2018 (N = 166)
Topic a	Library activities	Information technology	Multimedia libraries	Information technology	Information literacy	Online exhibition	Knowledge organization	Scientometrics
Words	Desidoc	Color	Multimedia	Library	Information	User	Search	Research
	Library	Technology	Digital	Cd	Library	Digital	Information	University
	Drdo	Printing	Http	Libraries	Libraries	Online	Subject	Papers
	Research	Paper	WWW	Information	Internet	Web	Data	India
	Libraries	Printers	Classification	Services	Education	WWW	Number	Publications
Topic b	Information technology	Library services	Information technology	Knowledge organization	Library management system	Open access	User studies	Citation analysis
Words	Information	Information	Databases	Data	Knowledge	Access	Information	Journals
	System	Services	Information	Database	System	Journals	Students	Journal
	Data	Libraries	Cd	Library	Data	Research	Library	Citation
	Systems	Document	Database	Information	Network	Open	Learning	Articles
	Cd	Service	Science	Entity	Management	Resources	Respondents	Number
Topic c	Information organization	Information system	Information organization	Knowledge management system	Information retrieval system	Library education	Scientometrics	Digital libraries
Words	Science	Systems	Library	Information	Search	University	Research	Search
	Defence	Computer	Libraries	Knowledge	Web	Library	Papers	Digital
	Information	Work	Electronic	Management	Library	Education	University	Information
	Shri	System	Collection	Technology	Text	Lis	Journal	System
	Delhi	Text	Access	Software	Digital	Science	Journals	Patent

Table 4.1 (continued)

Labels	1981–1990 (N = 48)	1991–1995 (N = 46)	1996–1999 (N = 94)	2000–2003 (N = 55)	2004–2007 (N = 79)	2008–2011 (N = 202)	2012–2015 (N = 238)	2016–2018 (N = 166)
Topic d	Library services	Information literacy	Information system	ICT	Copyright	Bibliometrics	Library resources and services	Knowledge management system
Words	Electronic	Software	Information	Web	Copyright	Publications	Library	Information
	Document	Database	Systems	Search	Security	Research	Libraries	Library
	Documents	Information	Marketing	Page	Patents	Cent	Resources	Technology
	Paper	Online	Knowledge	Text	Protection	Papers	Services	Management
	Printing	User	Services	File	Act	India	Users	University
Topic e	Information system	Library automation	Types of libraries	Bibliometrics	Bibliometrics	Knowledge management system	Webometrics	Webometrics
Words	Computer	Library	University	Papers	Science	Information	Web	Library
	System	System	Research	Research	Desidoc	Library	University	
	Software	Science	Library	Materials	Research	University	Web	
	Translation	Cost	Libraries	India	Technology	Websites	Libraries	
	Work	Automation	Guide	University	Information	Technology	Table	
Topic f	ND*	ND*	Information retrieval system	ND*	Scientometrics	User studies	Open source software	ICT
Words			Data		Papers	Software	Students	Mobile
			System		Cent	Open	Social	
			Search		India	Http	Research	
			User		Research	Patent	Books	
			Software		Publications	Knowledge	Information	

Table 4.1 (continued)

Labels	1981–1990 (N = 48)	1991–1995 (N = 46)	1996–1999 (N = 94)	2000–2003 (N = 55)	2004–2007 (N = 79)	2008–2011 (N = 202)	2012–2015 (N = 238)	2016–2018 (N = 166)
Topic g	ND*	ND*	ND*	ND*	ND*	Information retrieval system	Public libraries	User studies
Words						Web Data System Systems Ontology	Information Development Social Public Education	Library Resources Users Libraries Information
Topic h	ND*	ND*	ND*	ND*	ND*	ND*	Virtual exhibition	ND*
Words							Digital Virtual Content User Preservation	
Topic i	ND*	ND*	ND*	ND*	ND*	ND*	E-Resources	ND*
Words							Books Http WWW Book Electronic	

Note: ND* stands for *Not Mentioned*

4B: RapidMiner

About the Case Study This case study has been adapted from Lamba and Madhusudhan [21] to illustrate a topic modeling problem using RapidMiner and provides a step-by-step tutorial to solve it.

Problem "Electronic Theses and Dissertations (ETDs) are one of the most frequent type of educational resource" [21] that is being used by the academicians. As the number of ETDs is increasing every day, "the issue of organizing, managing, searching, and disseminating information from the ETDs has gained attention" [21] to enhance the searching and information retrieval performance of the ETD databases.

Dataset A folder containing plain text files[105] for 441 electronic theses and dissertations (ETDs) retrieved from ProQuest Dissertations and Theses (PQDT) Global database from 2014 to 2018 for the library science subject.

About the Tool Refer to Chap. 10, Sect. 10.2.3, to know more about `RapidMiner`.

Theory In `RapidMiner`, `Latent Dirichlet Allocation (LDA)` identifies topics using the `ParallelTopicModel` of the `Mallet` library with `SparseLDA` "sampling scheme and data structure" [21] in addition to `Gibbs Sampling`. LDA provides topic diagnostics in `RapidMiner` depending on the number of top words and can be tweaked using available additional parameters.

Methodology Figure 4.7 shows the workflow used to perform the topic modeling of documents in RapidMiner.

The following screenshots demonstrate the steps which were taken to perform topic modeling in the `RapidMiner` platform:
Step 1: "Read Document" is a sub-operator which is nested inside the "Loop Files" operator. The folder containing the sample plain text files was first to read into the sub-operator (Fig. 4.8).
Step 2: Various text pre-processing operators that performed tokenization, transforming of words to lower cases, filtering of stopwords, stemming, generating n-grams, and filtering the tokens by length were nested inside the "Loop Collection" operator. Each file then went through every text pre-processing operator in a loop-like manner until all got processed (Fig. 4.9).
Step 3: The "play" button was pressed once the authors were satisfied with the selection of parameters in the "Extract Topics from Documents (LDA)" operator (Fig. 4.10).

---

[105] https://github.com/textmining-infopros/chapter4/tree/master/4b_dataset.

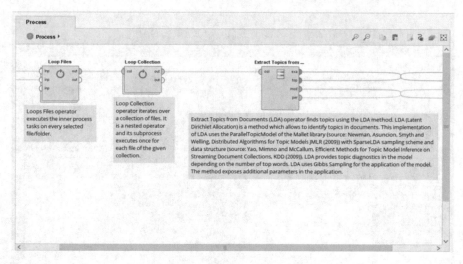

Fig. 4.7 Screenshot showing workflow for topic modeling in RapidMiner

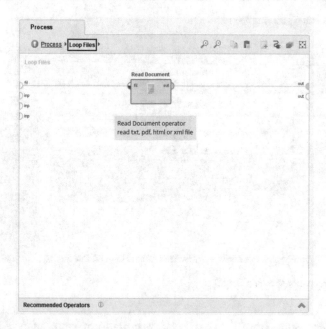

Fig. 4.8 Screenshot showing Step 1

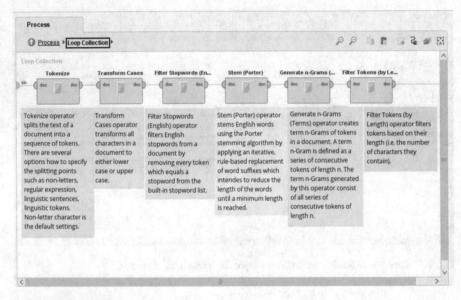

Fig. 4.9 Screenshot showing Step 2

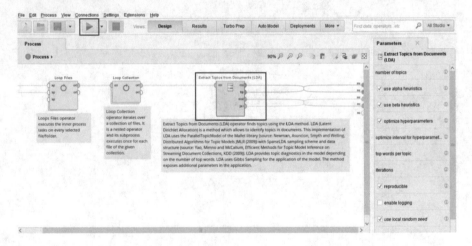

Fig. 4.10 Screenshot showing Step 3

Results The results from the RapidMiner platform were analyzed to assign appropriate topics to the corpus of ETDs. Table 4.2 sums up the "LDA results for the study that shows the labeling of the topics (a through h) in descending order as per their probability values wherein Topic a had the highest probability value. In order to determine the topics, both co-occurring words and representative

Table 4.2 Latent Dirichlet Allocation topic and word result for PQDT Global ETDs during 2014–2018 (©2020 Cadernos BAD, all rights reserved—reprinted under Creative Commons CC-BY license, published in Lamba and Madhusudhan [21])

Topic a *Children literature*	Topic b *Academic library*	Topic c *Information retrieval*	Topic d *Archival science*	Topic e *User study*	Topic f *Digital library*	Topic g *Library leadership*	Topic h *Digital communication*
Book	Student	Inform	Archiv	Inform	Data	Librari	Digit
Read	School	Search	Collect	Particip	Research	Servic	Commun
Librari	Librarian	User	Histori	Research	Journal	Librarian	Archiv
Children	Inform	System	Book	Studi	Scienc	Leadership	Media
School	Research	Task	Record	Social	Access	Staff	Technolog

ETDs[106] were taken into account. Representative ETDs are the top five ETDs that were ranked according to the highest topic proportion value for a chosen topic" [21].

We identified "eight core topics where Number of articles=441; Number of Words=5; AlphaSum=1.874; Beta=0.06. The evidence from high-loading keywords and most representative ETDs demonstrated that `Topic a` was about *children literature* with attention to reading habits. `Topic b` was about *academic library* with a focus on school library, research, information, librarians, and students" [21]. Further, "while `Topic c` was about *information retrieval* with an emphasis on searching, user, system, and task; `Topic d` showed a focus on *archival science* with an emphasis on collection, records, and history as opposed to `Topic e` which was centered around *user study* with an emphasis on information, participants, research, and society" [21]. Furthermore, "`Topic f` showed a focus on *digital library* with an emphasis on data, research, journal, and access in contrast with `Topic g` which was on *library leadership* with an emphasis on staff, librarian, and service in libraries. Lastly, `Topic h` was on *digital communication* with an emphasis on archives, media, and technology" [21].

4C: R

Problem If you have a particular set of documents such as research articles, electronic theses and dissertations, newspaper articles, archived librarian chats, or reference questions and answers for a specific discipline and want to tag them based on their topics/sub-domains.

Goal To identify patterns or groups of similar documents within a particular type of dataset of known discipline.

---

[106] https://github.com/textmining-infopros/chapter4/mass/ace/4b_supplementary.pdf.

Prerequisite Familiarity with R language.

Virtual RStudio Server You can reproduce the analysis in the cloud without having to install any software or downloading the data. The computational environment runs using BinderHub. Use the link (https://mybinder.org/v2/gh/textmining-infopros/chapter4/master?urlpath=rstudio) to open an interactive virtual RStudio environment for hands-on practice. In the virtual environment, open the `stm.R` file to perform topic modeling.

Virtual Jupyter Notebook You can reproduce the analysis in the cloud without having to install any software or downloading the data. The computational environment runs using BinderHub. Use the link (https://mybinder.org/v2/gh/textmining-infopros/chapter4/master?filepath=Case_Study_4C.ipynb) to open an interactive virtual Jupyter Notebook for hands-on practice.

Dataset A CSV file[107] containing metadata for 98 electronic theses and dissertations (ETDs) from Shodhganga database 2013–2017 in library and information science (LIS) subject.

About the Tool Refer to Chap. 10, Sect. 10.2.1, to know more about R.

Theory Structural topic modeling (STM) performs topic modeling at document-level metadata using "fast variational approximation. The `stm` package of R provides many useful features, including rich ways to explore topics, estimate uncertainty, and to visualize quantities of interest" [24].

Methodology and Results The libraries and the dataset required to perform topic modeling in R were loaded.

```
#Load libraries
library(tidyverse)
library(tidytext)
library(stm)
library(ggplot2)
library(RColorBrewer)

#Load dataset
data <- read.csv("https://raw.githubusercontent.com/
textmining-infopros/chapter4/master/4c_dataset.csv?
token=ARBWLQ3RCKTIK7PFAPDEUD3ACZJYA")
```

The loaded data was then cleaned to remove stopwords, punctuation, numbers, and whitespaces.

---

[107] https://github.com/textmining-infopros/chapter4/blob/master/4c_dataset.csv.

#Text pre-processing
```
processed <- textProcessor(data$Title,
removepunctuation = TRUE,
metadata = data)
```

```
out <- prepDocuments(processed$documents,processed$vocab,
processed$meta)
docs <- out$documents
vocab <- out$vocab
meta <- out$meta
```

Structural topic modeling (STM) was initialized with five topics (k).

#Running Structural Topic Modeling (STM)
```
STM <- stm(documents = out$documents,
vocab = out$vocab,
K = 5,
prevalence =~ Year,
max.em.its = 75,
data = out$meta,
init.type = "Spectral",
verbose = FALSE)
```

Figure 4.11 shows four different ways of representing the document-topic "proportion of the corpus that belongs to each topic" [24]. The interactive LDAVis, which is the output of *Method 4*, can be visualized at https://textmining-infopros. github.io/chapter4/. Further, the figures listed the top words with the highest probability that were associated with the selected number of topics.

#Different ways to plot topics with document-topic proportion
#Method 1: Plotting top words using stm package
```
plot(STM)
```

#Method 2: Plotting MAP histogram using stm package
```
plot(STM, type="hist")
```

(continued)

#Method 3: Visualizing topic model using ggplot2
```
topics <- tidy(STM, matrix = "beta")

  top_terms <- topics %>%
group_by(topic) %>%
top_n(10, beta) %>%
ungroup() %>%
arrange(topic, -beta)

  top_terms %>%
mutate(term = reorder(term, beta)) %>%
ggplot(aes(term, beta, fill = factor(topic))) +
geom_col(show.legend = FALSE) +
facet_wrap(~ topic, scales = "free") +
theme_minimal()+
theme(plot.title = element_text(hjust = 0.5, size = 18))+
labs(title="STM for Indian ETDs", caption="Top Terms")+
ylab("")+
xlab("")+
coord_flip()
```

#Method 4: Interactive Visualization
```
ldavis <- toLDAvis(STM, docs, R = 5,
plot.opts = list(xlab = "PC1", ylab ="PC2"),
lambda.step = 0.1,
out.dir = "LDAvis",
open.browser = interactive(),
as.gist = FALSE,
reorder.topics = TRUE)
```

In the following code, n represents the number of representative documents, and topics indicate the topic number. The code ran each time for all the five topics (where topics = 1, 2, 3, 4, 5) for a constant value of n (where n = 5). Table 4.3 presents the top five representative ETDs for the modeled topics, which are ranked according to their probability.

#Understanding topics through top 5 representative documents
```
findThoughts(STM, texts = data$Title, n = 5, topics = 5)
```

It can be observed that both Topic-5 and Topic-4 are more common in the corpus if compared to other topics (Fig. 4.11). The high probability keywords in Fig. 4.11 and the representative ETDs in Table 4.3 indicate that Topic-1

Fig. 4.11 Different ways to plot topics with document-topic proportion. (**a**) Output of Method 1. (**b**) Output of Method 2. (**c**) Output of Method 3. (**d**) Output of Method 4

(a)

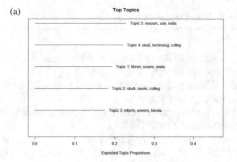

(b)

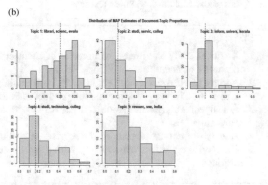

(c)

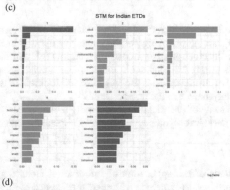

(d)

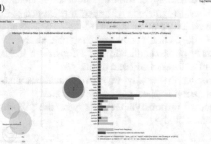

Table 4.3 Representative ETDs

Topic 1

1. Content analysis of library services in Library through library website with special reference to Mumbai Region: a study
2. Use of library classification schemes in the ICT environment in selected libraries in national capital region: a study
3. Performance evaluation and social-cultural contribution of public libraries in south region Ahmednagar district (M.S.)
4. A user study of chemical industry libraries in Maharashtra
5. Quality library services in the Department of Library and Information Science of universities of the Maharashtra state: a study

---

Topic 2

1. A critical study and review of libraries of engineering colleges and polytechnics in Beed district of Maharashtra
2. A study of NBA-accredited engineering college libraries in Maharashtra with relevance to marketing of library and information product sources and services
3. A study of library automation and networking in district public libraries of Maharashtra government
4. A study of information sources and services of architecture college libraries in Maharashtra
5. College library effectiveness study with special reference to the Tinsukia and Dibrugarh district

---

Topic 3

1. Information literacy of research scholars of universities in Kerala
2. Marketing orientation of the university libraries in Kerala in information dissemination
3. Intellectual property information and its role and importance in knowledge generation and industrial development in India
4. User awareness and use of electronic journals in technical university libraries of Punjab, Haryana, Chandigarh, and Delhi
5. A study of the scientific productivity and information use pattern of scientists in the context of new information technology with special reference to universities in Kerala

---

Topic 4

1. Study of the application of information technology in the treatment and preparation of medicine in Ayurveda with special reference to Kerala
2. Information communication technology skills among the library professionals of Engineering Colleges in Karnataka an analytical study
3. A study of the application of information technology in tribal medicine in Kerala with regard to forest medicinal plants
4. A comparative study of citation analysis among five online journals in library and information science: a scientometric study
5. Impact of information technology on the collection development in university libraries of Assam: a study

---

Table 4.3   (continued)

Topic 5

1. Information-seeking habits of software professionals in western India
2. Information literacy of management students in Mumbai metropolitan area
3. Design and development of collaborative model of health information literacy in Jammu Division
4. Development of prototype model of centralized library system for Sinhgad Institutes Higher Education Libraries using open-source software for enhancing library services
5. Design and development of network-based model for management of college libraries in Pune city with special reference to network securities

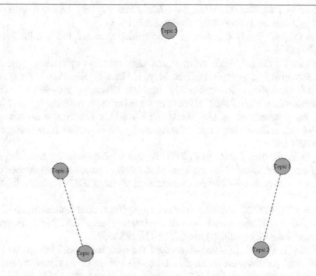

Fig. 4.12   Topic correlation network

focuses on *evaluation* specifically of resources, websites, and content; Topic-2 emphasizes on *library services*; Topic-3 represents *information management*; Topic-4 is about *information technology*; Topic-5 is on *user studies*.

Additionally, we identified the correlation between the topics using a network graph (Fig. 4.12). It was found that Topic-1 (*evaluation*) was related to Topic-2 (*library services*), and Topic-3 (*information management*) was related to Topic-4 (*information technology*). Topic-5 (*user studies*) was isolated and not related to other topics.

```
#Topic correlation
topicor <- topicCorr(STM)
plot(topicor)
```

References

1. Lamba M, Madhusudhan M (2019) Mapping of topics in DESIDOC. J Libr Inf Technol India Study. Scientometrics 120:477–505. https://doi.org/10.1007/s11192-019-03137-5
2. Lamba M, Madhusdhuan M (2018) Application of topic mining and prediction modeling tools for library and information science journals. In: Library practices in digital era: Festschrift in Honor of Prof. V Vishwa Mohan. BS Publications, Hyderabad, pp 395–401. https://doi.org/10.5281/zenodo.1298739
3. Blei DM (2012) Topic modeling and digital humanities. J Digital Humanit 2(1). https://journalofdigitalhumanities.org/2-1/topic-modeling-and-digital-humanities-by-david-m-blei/
4. Gelman A, Shalizi CR (2013) Philosophy and the practice of Bayesian statistics: Philosophy and the practice of Bayesian statistics. Br J Math Stat Psychol 66:8–38. https://www.stat.columbia.edu/~gelman/research/unpublished/philosophy.pdf
5. Alghamdi R, Alfalqi K (2015) A survey of topic modeling in text mining. Int J Adv Comput Sci Appl 6:147–153
6. Pfeifer D, Leidner JL (2019) Topic grouper: An agglomerative clustering approach to topic modeling. In: Azzopardi L, Stein B, Fuhr N, Mayr P, Hauff C, Hiemstra D (eds) Advances in information retrieval. Springer International Publishing, Cham, pp 590–603
7. Chuang J, Manning CD, Heer J (2012) Termite: visualization techniques for assessing textual topic models. In: Proceedings of the international working conference on advanced visual interfaces - AVI '12. ACM Press, Capri Island, Italy, p 74. https://idl.cs.washington.edu/files/2012-Termite-AVI.pdf
8. Chuang J, Gupta S, Manning CD, Heer J (2013) Topic model diagnostics: assessing domain relevance via topical alignment. In: Proceedings of the 30th international conference on machine learning, Atlanta, Georgia, USA. https://vis.stanford.edu/files/2013-TopicModelDiagnostics-ICML.pdf
9. Murdock J, Allen C (2015) Visualization techniques for topic model checking. In: Proceedings of the twenty-ninth AAAI conference on artificial intelligence. AAAI Press, Austin, Texas, pp 4284–4285. https://dl.acm.org/doi/10.5555/2888116.2888368
10. Sievert C, Shirley K (2014) LDAvis: A method for visualizing and interpreting topics. In: Proceedings of the workshop on interactive language learning, visualization, and interfaces. Association for Computational Linguistics, Baltimore, Maryland, USA, pp 63–70. https://www.aclweb.org/anthology/W14-3110.pdf
11. Yang Y, Yao Q, Qu H (2017) VISTopic: A visual analytics system for making sense of large document collections using hierarchical topic modeling. Visual Informatics 1:40–47. https://doi.org/10.1016/j.visinf.2017.01.005
12. Gobbo B, Balsamo D, Mauri M, Bajardi P, Panisson A, Ciuccarelli P (2019) Topic tomographies (TopTom): A visual approach to distill information from media streams. Comput Graph Forum 38:609–621. https://doi.org/10.1111/cgf.13714
13. Karpovich S, Smirnov A, Teslya N, Grigorev A (2017) Topic model visualization with IPython. In: 2017 20th conference of open innovations association (FRUCT). IEEE, St-Petersburg, Russia, pp 131–137. https://www.fruct.org/publications/fruct20/files/Kar2.pdf
14. Smith A, Hawes T, Myers M (2014) Hierarchie: Interactive visualization for hierarchical topic models. In: Proceedings of the workshop on interactive language learning, visualization, and interfaces. Association for Computational Linguistics, Baltimore, Maryland, USA, pp 71–78. https://nlp.stanford.edu/events/illvi2014/papers/smith-illvi2014b.pdf
15. Liu L, Tang L, Dong W, Yao S, Zhou W (2016) An overview of topic modeling and its current applications in bioinformatics. SpringerPlus 5(1):1608. https://doi.org/10.1186/s40064-016-3252-8.
16. Blei DM, Ng AY, Jordan MI (2003) Latent dirichlet allocation. J Mach Learn Res 3:993–1022. https://www.cs.columbia.edu/~blei/papers/BleiNgJordan2003.pdf

17. Lamba M, Madhusudhan M (2018) Metadata tagging of library and information science theses: Shodhganga (2013–2017). In: Beyond the boundaries of rims and oceans globalizing knowledge with ETDs - ETD 2018. Taipei, Taiwan. https://doi.org/10.5281/zenodo.1475795

18. Mehler A, Waltinger U (2009) Enhancing document modeling by means of open topic models: Crossing the frontier of classification schemes in digital libraries by example of the DDC. Library Hi Tech 27:520–539. https://doi.org/10.1108/07378830911007646

19. Angus V (2016) What's it all about? Indexing my corpus using LDA. In: Seen Another Way. https://seenanotherway.com/whats-it-all-about-indexing-my-corpus-using-lda/. Accessed 31 Oct 2020

20. Ayadi R, Maraoui M, Zrigui M (2014) Latent topic model for indexing arabic documents. Int J Inf Retriev Res 4:29–45. https://doi.org/10.4018/ijirr.2014010102

21. Lamba M, Madhusudhan M (2020) Mapping of ETDs in ProQuest dissertations and theses (PQDT) global database (2014–2018). Cadernos BAD 2019(1):169–182. https://www.bad.pt/publicacoes/index.php/cadernos/article/view/2034

22. Lamba M (2019) Text analysis of ETDs in ProQuest dissertations and theses (PQDT) Global (2016–2018). In: Digital transformation for an agile environment - ICDL 2019, New Delhi. https://doi.org/10.5281/zenodo.3545907

23. Carlson J, Harris K (2020) Quantifying and contextualizing the impact of bioRxiv preprints through automated social media audience segmentation. PLOS Biology 18:e3000860. https://doi.org/10.1371/journal.pbio.3000860

24. Roberts ME, Stewart BM, Tingley D (2019) stm: R package for structural topic models. J Stat Softw 91(2):1–41. https://cran.r-project.org/web/packages/stm/vignettes/stmVignette.pdf

Additional Resources

1. https://sicss.io/2020/materials/day3-text-analysis/topic-modeling/rmarkdown/Topic_Modeling.html

2. https://tedunderwood.com/2012/04/07/topic-modeling-made-just-simple-enough/

3. https://monkeylearn.com/blog/introduction-to-topic-modeling/

4. https://www.scottbot.net/HIAL/index.html@p=221.html

5. https://journalofdigitalhumanities.org/2-1/topic-modeling-a-basic-introduction-by-megan-r-brett/#topic-modeling-a-basic-introduction-by-megan-r-brett-n-2

6. https://thesai.org/Downloads/Volume6No1/Paper_21-A_Survey_of_Topic_Modeling_in_Text_Mining.pdf

7. https://journalofdigitalhumanities.org/2-1/topic-modeling-and-figurative-language-by-lisa-m-rhody/

8. https://journalofdigitalhumanities.org/2-1/words-alone-by-benjamin-m-schmidt/

9. https://journalofdigitalhumanities.org/2-1/review-papermachines-by-adam-crymble/

10. https://journalofdigitalhumanities.org/2-1/review-mallet-by-ian-milligan-and-shawn-graham/

11. https://www.cs.columbia.edu/~blei/papers/Blei2012.pdf

12. https://www.gotostage.com/channel/f12ef6e05ed24e10b48a53c70a39e266/recording/1141b5-1dc756442998995d0e7fd3b89d/watch

Chapter 5
Network Text Analysis

Abstract This chapter covers the theoretical framework for network text analysis, including its advantages, disadvantages, and various essential features. Further, it covers various open-source tools that can be used to make a text network. Information professionals may use network text analysis to answer various research questions and get a better visual representation of textual data. Use cases that show the application of network text analysis in libraries are also covered. Lastly, to demonstrate the application of network text analysis in libraries better, two case studies are performed using the bibliometrix and textnets packages in R language.

5.1 What Is Network Text Analysis?

Network analysis is a technique that depicts the association between units of analysis where networks are mathematical objects that are composed of nodes (or vertices) and are connected by edges (or links or arcs). Nodes are generally people and edges in a social network that describe some kind of social relationships between people, such as affiliations and friendships. In social sciences, social networks describe the different patterns of clustering and determine a person's position within or between those clusters. Thus, network analysis describes relationships between people, but some early pioneers noticed that it could also portray association between words. In 2005, Diesner and Carley [1] coined the phrase network text analysis (NTA) "to describe a wide variety of computer-supported solutions that enable analysts to extract networks of concepts from texts and to discern the meaning represented or encoded within" [2]. For instance, a collection of texts may be presented as a text network where every node represents a text, and the edges show the similarities between the words used in two or more documents. A text network can also be used to visualize individual words as nodes and their frequency with which they co-occur in the texts as the edges (Fig. 5.1).

Measuring word co-occurrence in large corpora, texts, sentences, or small chunks of texts for different models and approaches is a standard procedure in the field of computer linguistics, where these *bag-of-words* approaches ignore the structure of

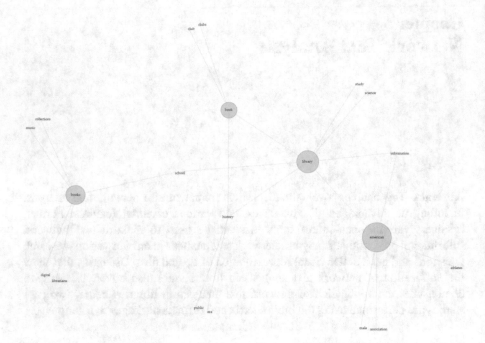

Fig. 5.1 Example of a text network

the text. Many recent studies [3–7] have started studying network structures in texts in more depth. Various types of networks can be created from the text:

1. Semantic Networks: It connects concepts that involve human decisions on what concepts are connected to other concepts. Thus, it usually aims to represent knowledge about some domain in a way that simulates human mental representations.
2. Word Association Networks: It connects target words that come up in subjects' mind when reading a cue.
3. Word Co-occurrence Network: It connects two words that are present closely together in a text and are linked by an edge.

In network text analysis (NTA), a document is depicted as a network of words with key concepts that are related to it. NTA focuses on the topology of networks and their structure in order to gain interesting insights about the text. It is different from semantic text analysis, which focuses on interpreting the relation between the terms based on their meaning. It "can be used to extract the underlying social and organizational structure from texts in an effective and efficient way where texts can be represented as networks of concepts (such as ideas, people, resources, organizations, events, etc.) and the connections between them" [1]. Some of the crucial assumptions for NTA are:

1. Embodied *"language and knowledge* in the text can be modeled as a network of words and relations between them" [2].
2. "position of concepts within the network provides insights into the meaning or prominent themes of the text as a whole" [2].

We can use the fundamental principle of network graph theory and apply it in text networks by (i) drawing edges between two people based on their similarities and the types of words they use when they write, talk, speak, or work on, and (ii) treating the words as the nodes of the network and drawing edges between the words when many people use the same group of words. At root, text networks perform similarly to topic models by looking for patterns and grouping of words that you might call topics. However, it can also be used to cluster authors of the documents themselves or group them according to their similarities. Social networks have valence on their edges, for example, by integrating sentiment analysis with sentence parsing in order to connect authors who talked about the same things in similar ways and connect words to each other by how they are used in terms of their sentiment by multiple authors. Thus, authors can have positive or negative relationships with each other. Once we identify such network structure, we can build theories of human relationships such as balance theory.

5.1.1 Two-Mode Networks

Opsahl [8], in his paper, described that "most networks are one-node networks with one set of nodes that are similar to each other. However, several networks are two-mode networks (also known as bipartite networks) with two different sets of nodes and edges occurring only between nodes belonging to different sets." He further emphasized that "often a distinction is made between the two sets of node by determining which set is more responsible for edge creation (primary or top node set) than the other (secondary or bottom node set). Some of the examples of two-mode networks include (i) attendance of a group of women (primary node set) to a series of events (secondary node set) where a woman would be linked to an event if she attended it" [8]; (ii) scientific collaboration networks where both scientists and papers are "the two sets of nodes and a scientist (primary node set) is linked to a paper (secondary node set) if he or she is listed as the author of that paper" [8]; and (iii) word-based projection of every word that co-occurs across different presidential speeches that are laid out in a network space using a community detection algorithm (i.e., Louvain Community Detection method) where the words are placed nearby to the words that are similar in terms of their connection to others. Thus, you get a network structure between words that have some similar concept/theme/subject.

Two-mode networks (Fig. 5.2) are usually transformed into a one-mode network before commencing any analysis "because most network measures are defined for one-mode networks" [8] (Fig. 5.3). Thus, in a two-mode text network (Fig. 5.2), "one node set will always be comprised of the words found in the documents

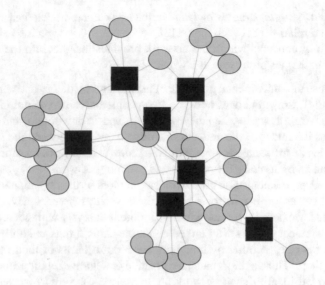

Fig. 5.2 Diagram showing an example of a two-mode network

analyzed (Fig. 5.3a), and the other node set can be the documents themselves, or some other type of metadata about those documents such as the author's name or the date when the document was published or created" [9] (Fig. 5.3b).

Similar to topic modeling, you can cluster words in latent themes or topics in the form of word-based projections of the text network. Figure 5.2 shows an example of a two-mode network where the nouns/compound nouns/phrases are presented in yellow circles, whereas the authors of those words are presented in red squares. Figure 5.3a represents a word network, whereas Fig. 5.3b presents an author network. In this example, weaker edges can be used to represent similar commonly used words by the authors in contrast to stronger edges that can be used to show the usage of similar unusual words based on TF-IDF by the authors. These edges are created with weights that are the sum of the overlapping TF-IDF of terms that any two authors have used in a text corpus. So, every author (red square) has a connection to a noun (yellow circle) that is used by them and thus can be used to describe a pattern where the authors jointly use similar words.

5.1.2 Centrality Measures

A network can be visualized at three different levels: **network** (highest-level overview that reveal patterns like overall connectivity and balance in a network), **group** (when a graph is viewed at a sub-network level, one segment at a time, and reveals clusters of connections, it is easier to see the dynamics between groups), and **entity** (the node-by-node level where small clusters are identified

Fig. 5.3 Diagram showing the transformation of a two-mode network into a one-node network. (**a**) First node set representing words. (**b**) Second node set representing documents or metadata

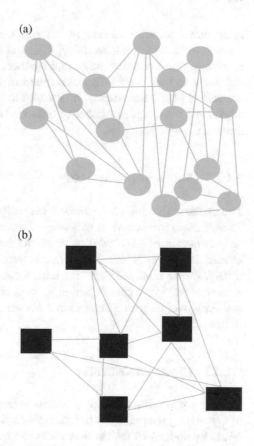

and individual entities within their connected environment are studied). A node's centrality measures its structural importance or prominence in a network, where a high centrality score could indicate power, influence, control, or status. It is crucial to identify the *central* node in a network as it can be used to answer many research questions such as how people in between cluster of network often serve key roles and spread ideas or often receive disproportionately large social advantage by virtue of being able to connect to different group of people. It helps to look at who is in between discursive communities (*key broker* connecting the flow of ideas across community), or in the case of words, one can find the central words to the discourse that link different themes together and how different actors use those words by comparing the author projection to the word projection. We will now discuss a range of centrality measures to identify the most influential nodes in a text network.

5.1.2.1 Degree Centrality

Degree centrality measures nodes (in our case, words) with the most edges to other nodes by assigning importance scores in a text network. It can tell us "how many direct connections each node has to the other in a network. It is the simplest measure of a node's connectivity" [10] and helps to answers questions like "who is the most/least popular author in the network?"; "who is most connected?"; "who are likely to hold most information; and who can quickly connect with the wider network?" [10]. "Sometimes, it is useful to look at in-degree (number of inbound edges) and out-degree (number of outbound edges) as distinct measure" [11] for a particular set of problems.

5.1.2.2   Betweenness Centrality

The nodes with a high betweenness centrality act as *bridges* between other nodes. Thus, forming the shortest pathway of communication within a network helps to answer questions like "who or what can most strongly control information flow around the network?" and "who or what would cause the most disruption to the flow if they/it were removed?". It "is useful for analyzing communication dynamics but should be used with care as a high betweenness count could indicate someone hold authority over disparate clusters in a network or just that they are on the periphery of both clusters" [10].

5.1.2.3   Closeness Centrality

Closeness centrality measure is similar to betweenness centrality. However, instead of calculating the paths through each node, it calculates a node's proximity to other nodes by finding all the shortest paths in a network and then assigning "each node a score based on the sum of its shortest paths" [10]. Thus, it scores "each node based on their *closeness* to all other nodes in a network" [10]. It is most insightful when a network is a sparsely connected network as in a highly connected network; it will give a similar score to all nodes. Therefore, it can be used to find influencers in individual clusters. It can help answers questions like "who can most efficiently obtain information on other nodes in a network?" and "who could most quickly spread information in a network?".

5.1.2.4   EigenCentrality

EigenCentrality is similar to degree centrality but it considers the number of edges each node has, the number of edges their connections have, and so on throughout the network. "A node may have a high degree score (i.e., many connections) but a relatively low EigenCentrality score if many of those connections are with similarly low-scored nodes" [10] and vice versa. It helps answer questions like "who or what holds wide-reaching influence in a network?" and "who or what is essential in a network on a macro scale?".

5.1.2.5   PageRank

"PageRank is one of the famous ranking algorithms that is behind the Google search engine. Similar to EigenCentrality, PageRank uncovers influential or important nodes whose reach extends beyond just their connections" [10]. It identifies the significance of nodes by assigning a score based on the number of incoming edges (in-degree) by taking edge "direction and weight into account, therefore, edges can only pass influence in one direction and can pass different amounts of influence" [11]. Moreover, it is helpful in comprehending authority and citations. It helps answering questions like "who or what holds wide-reaching influence in a network?" and "who or what is important in a network on a macro scale?".

5.1.3   Graph Algorithms

In addition to centrality measures, it is essential to understand some of the crucial algorithms used to perform NTA. The significant difference between *undirected graphs* and *directed graphs* is that edges connect two nodes symmetrically in *undirected graphs* and asymmetrically in *directed graphs*. Some of the vital graph algorithms are:

1. Distance/Shortest Path: The *distance* function measures how the distance between the shortest path and two nodes in a network highlights the path that passes through the lowest number of nodes. Weight can be added to calculate the actual distances as well as the number of nodes.
2. Community Finding and Clustering: Clustering uncovers the community structure in large datasets and is based on modularity (a way of measuring how readily a network can be divided into modules or sub-networks). These algorithms help to find communities easily and help to understand more about the studied organizations.
3. Network Layouts: These algorithms decide where nodes are placed on the screen and are crucial in network visualization such as force-directed, circle, fruchterman and reingold, kamada and kawai, sphere, and star.

5.1.4   Comparison of Network Text Analysis with Others

Clusters cannot be detected visually or quantitatively within the text using keyword frequency. In contrast, tag clouds only tell about the relation of a specific word to the whole text but fail to tell about the relation between the words. Text networks help to identify the text's main context by providing insight about the dominating topics/themes/concepts within several embedded communities or clusters. The various advantages of NTA to automated text analysis approach [9] are as follows:

1. It helps to visualize the entire corpus in one picture.
2. It helps to see how dense the clusters of words are that are coherent to the topics. In contrast, topic models give a mixed membership model that allows accounting for words that are in between clusters, and we cannot see them or see how they guide the assignment of individual words into clusters. Thus, text networks help to visualize which clusters are the most cohesive and which words kind of bridge multiple clusters as very often we want to know about those kinds of in-between spaces and theories.
3. By understanding the "patterns of connections between words, one can identify their meaning in a precise manner compared to *bag-of-words* approach" [9].
4. The number of topics is chosen by default by the Louvain Community Detection method by NTA packages like `Textnets`. In contrast, in topic models, goodness-of-fit measures are taken to find the number of topics for the model. The goodness-of-fit measures guess the best number of topics that may fit the model and thus are not very reliable as they can result in shady groupings of words.
5. It "can be built out of documents of any length whereas topic models process short texts (such as tweets) very poorly" [9].
6. It is a "more sophisticated set of techniques to identify clusters within social networks than those employed in automated text analysis techniques" [9].

5.1.5   How to Perform Network Text Analysis?

The basic steps which are involved in performing NTA are:

1. Pre-processing of textual data covered in Sect. 3.3.
2. Converting the data to a graph and finding:

    a. "the most influential nodes (words) that function as a junction of circulation" [12]
    b. "the contextual clusters or distinct word communities (or themes) presented in the text" [12]
    c. "the main quantitative properties of the network" [12]

3. By using the information from the above step, which explores the relation "between the communities (contextual clusters) and the role of the junctions that link the clusters" [12]. It will help to identify the pathway of circulation in the text.
4. Finding alternative pathways to "exemplify the main agenda, but in different terms" [12]
5. Interpreting the results

5.1.6 Available Tools and Packages

There are many open-source tools/applications available for non-programmers to perform NTA and visualize the networks, such as `Wordij`,[1] `NodeXL`,[2] `Polinode`,[3] `VoyantTools`,[4] `VOSviewer`,[5] `Science of Science (Sci2)`,[6] `RapidMiner`,[7] `Orange`,[8] `LancsBox`,[9] `ConText`,[10] `Gephi`,[11] `Tableau`,[12] `Microsoft Power BI`,[13] and `Palladio`.[14] Though there are many network analysis packages in different programming languages and environment for programmers, some of the popular packages that can be used to perform network text analysis or can be used to visualize text networks are `textnets`,[15] `igraph`,[16] and `bibliometrix`[17] in R and `networkx`,[18] `pyvis`,[19] and `textnets`[20] in Python. Most of the tools mentioned above are covered in detail in Chap. 10.

5.1.7 Applications

Various applications of NTA include [12]:

1. It helps to determine the main context of the text quickly.
2. It helps to open the potential to interpret the text in a nonlinear fashion which will not be uncovered via standard sequential reading.

---

[1] https://www.wordij.net/download.html.

[2] https://nodexlgraphgallery.org/Pages/AboutNodeXL.aspx.

[3] https://www.polinode.com/.

[4] https://voyant-tools.org/.

[5] https://www.vosviewer.com/.

[6] https://cns.iu.indiana.edu.

[7] https://rapidminer.com/.

[8] https://orangedatamining.com/.

[9] https://corpora.lancs.ac.uk/lancsbox/download.php.

[10] https://context.lis.illinois.edu/download.php.

[11] https://gephi.org/users/download/.

[12] https://public.tableau.com/en-us/s/download.

[13] https://powerbi.microsoft.com/en-us/downloads/.

[14] https://hdlab.stanford.edu/palladio/.

[15] https://github.com/cbail/textnets.

[16] https://igraph.org/r/.

[17] https://bibliometrix.org/.

[18] https://pypi.org/project/networkx/.

[19] https://pyvis.readthedocs.io/en/latest/tutorial.html#example-visualizing-a-game-of-thrones-character-network.

[20] https://pypi.org/project/textnets/.

3. It helps to plot a story from the text clearly and helps to uncover the underlying psychological narrative of the text.
4. It can help to perform group sentiment analysis where interviews can be conducted "with the group members to bring together the resulting graphs to reveal the key and peripheral concepts of the group" [12].
5. It helps to reveal potential areas for group collaborations.
6. It can be used to compare differences or similarity between different types of textual data.
7. It is a way to present text visually and "can be very useful for writers, editors, and copywriters as it can allow for the more holistic perception of its interconnection and open up more possibilities for interpretations" [12].
8. It can allow an easy-to-use interface with a "whole range of navigational, archival and searching tools for textual data" [12] if translated to a tool/website/product as it focuses on the interconnectivity between different concepts, texts, and contextual data.
9. It can allow "dynamic analysis of textual data where the graph can be visualized in spatio-temporal frame" [12] and can be used to create an interface for navigating related content.
10. It can be used to create an interface for a "live audio-visual cross-content navigational system where speech recognition mechanism could provide an immediate graphical representation of the speech and show how it can be related to other audio-visual content both within the same context and outside it" [12].

5.1.8   Advantages

There are several advantages of NTA [13, 14]:

1. The text can be seen at once instead of waiting for it to unfold with time. "Visualizing text as a network removes the variable of time" [12]. Thus, the projections can be analyzed over time.
2. It helps to trace the pathways for meaning circulation within the text.
3. It goes beyond the traditional text analysis tools (such as tag clouds, finding the most frequently mentioned concept) and helps to analyze actual relations and the processes that align the words in a specific way rather than the terms themselves.
4. It unlocks the potentiality of a narrative, i.e., it offers many responsibilities for interpretation and reading which a dominant narrative would generally suppress.
5. It is less computational-heavy compared to other analyses.
6. It helps "to explain a range of outcomes and understanding patterns of connections between words which helps to identify their meaning in a more precise manner than the *bag of words* approach" [9].
7. It "can be built out of documents of any length" [9].
8. It is a "more sophisticated set of techniques for identifying clusters within social networks" [9].

5.1.9 Limitations

Some of the significant limitations of NTA are:

1. The algorithms are lengthy and may take a longer time to process.
2. It focuses on the *bigger picture* and can neglect more minor instances that play an essential role in specific analysis.
3. Like topic modeling, text network depends on the interpretability of the analyzer, i.e., how he/she interprets the results.
4. In some cases, the overlapping of too many nodes can make it difficult to read the network, especially in very dense networks.

5.2 Topic Maps

Topic maps are conventional reference models that can be used to apply familiar techniques from the library and information science domain, such as cataloging, or indexing. Initially, they were used to represent knowledge structures intrinsic to printed book indexes to answer the problems of managing the information. Later, they were extended to illustrate subject-based classification and metadata information by reusing the current classification schemes and methods [15]. They are a comparatively novel approach that can better organize websites than current information organization techniques and present a method that includes traditional methods of indexing and information representation, with cutting-edge techniques of linked data [15]. They can signify the interrelation between various metadata fields and link them to the corresponding information resource in a digital library. They use a Web 3.0 enabling technology and a kind of semantic-web technology similar to RDF. Its standards and specifications are designed to gather all the information about a subject at a single location. It supports several query languages, modeling languages, and file formats. It has an international industrial standard (ISO 13250) focusing on finding the information for information management [16].

Topic maps were originated during the integration of electronic indexes in a digital environment and subject-based classification techniques and are commonly used for web portal development, content management, and knowledge management in LIS. These maps are generally structured around *topics*, where every *topic* is used to denote an entity. *Topics* represent concepts (which are known as *subjects*) in a similar way as terms are referred to as concepts in an indexing language [15]. A topic map consists of three constructs, viz., occurrences, names, and associations, that describe the subjects signified by the topics (Fig. 5.4). These constructs define the name, property, and relationship of metadata information for different information resources. Thus, a topic map presents metadata as:

- *Topics*, where each topic represents concepts such as countries, people, individual files, software modules, events, or organizations

Fig. 5.4 Diagram showing a
topic map

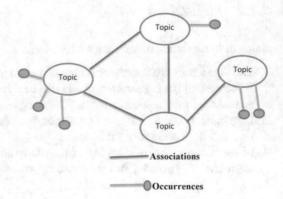

- *Association*, where specific relationships are identified between the topics
- *Occurrences*, where different library resources related to a specific topic are identified

Using topic maps, users can visualize a multidimensional topic space of knowledge instead of going through each resource from the enormous data one by one. Therefore, they can determine which information resources are relevant to them. Further, the process of querying in topic maps can be much more precise and faster than performing a full-text searching. Indexes that are based on topic maps become the prerequisite for information delivery and processing. The proximity of topic maps with semantic networks gives an impression of how topic maps can become valuable resources to information professionals in the digital era. For instance, Wlodarczyk [16] used topic maps to visualize the subject headings and complex structure of terms as a whole for the National Library of Poland. Tools like Ontopoly,[21] Topic Maps Toolbox,[22] tinyTIM,[23] and TM4J,[24] can be used to build topic maps.

5.2.1   Constructs of Topic Maps

1. **Topics**: A topic can be a person, entity, or concept that represents a subject.
2. **Subjects**: The subject corresponds to an idea. A topic can consist of any quantity of names. If several names are given to a topic, then all the names denote the same subject, wherein all the names are substitutes of each other. The same name can be given to different topics, which is generally not seen in the context of taxonomies and thesauri.

---

[21] https://ontopia.net/doc/current/ontopoly/user-guide.html.

[22] https://www-edc.eng.cam.ac.uk/cam/documentation/Topic%20map%20toolbox.

[23] https://sourceforge.net/projects/tinytim/.

[24] https://tm4j.org/.

3. **Types**: Topics can be characterized according to their kind and can be typed to distinguish which of them are taken and which are not. This capability is absent from traditional library classification methods.
4. **Names**: Generally, topics have exact names, but it is not compulsory, e.g., in cross-referencing of pages where a relation to a topic has an uncertain name. The ISO standard for topic maps provides a facility to assign multiple *base names* and *variants* of each base name for a specific use.
5. **Occurrences**: A topic may be related to one or more library resources such as webpages, videos, images, dissertations, newspapers, or ebooks that are essential to the topic. Such library resources are called the occurrences of the topic. Occurrences relate topics to related resources and indicate where one can find information about the subject similar to the function performed by page numbers or subject property in a book index and Dublin Core metadata schema. Here, an object is associated with the subjects. Thus, occurrences connect topics to different library resources that consist of information about them. They have *types* that distinguish different kinds of relationships and can differentiate maps from portraits or video clips from an audio clip. Also, a *scope* can be applied to an occurrence, for instance, to decide whether resources are suitable for bachelor's students or master's students. Moreover, occurrences use URIs to find the resource linked to the topic or string of information saved in the topic map with properties such as description, date, or page number. As *occurrence types* are topics, the creator of the topic map can define *occurrence types* at will, whereas in traditional library science techniques, open vocabulary is not allowed for properties and occurrences.
6. **Associations**: Associations demonstrate relationships between subjects and occurrences in topic maps, which can further be typed. They are inherently multidirectional, i.e., if B is related to A, then A is also related to B. Associations can express any kind of relationship and are very different to traditional classification schemes as (i) they show network instead of hierarchies for subjects, (ii) they show defined relationships between the subjects, and (iii) they allow complex query searches [17].
7. **Association types**: Association types cluster topics that have a similar relationship to any selected topic and help to provide easy-to-use interfaces for navigating a large corpus of resources.
8. **Association roles**: Every topic that contributes to association shows a vital role in that association called association role. Association roles can be a topic and can be typed.

5.2.2 Topic Map Software Architecture

Topic map software supports the simple conception of new topics and their associations and occurrences. A typical topic map tool allows browsing, import, manipulation, and export of existing or new maps with features like addition, deletion, and renaming that are offered for manipulating such features of the topic maps. The following are the four major features of topic map software:

1. **Designing and editing topic maps**: The creator of topic maps identifies the topics (topic types), occurrences (occurrence role types), and association (association types) for the information resources he/she wants in designing the topic map and writes the code in XTM syntax.
2. **Representation and navigation of topic maps**: For efficient navigation, the whole topic map is represented so that users can emphasize on any portion of the topic map and search for accurate information about any particular topic. Different representative features may be present in topic map software for users to navigate freely at a different level of detail.
3. **Visualization of topic maps**: Graphs and trees can be used to represent topic maps with nodes and edges of different shapes and colors.
4. **Querying topic maps**: Topic map data can be queried using the TOLOG language.

5.2.3   Typical Uses

1. Creating a topic map to index information resources and linking those information resources to other resources present in the library.
2. Subject maps are designed to amalgamate numerous research databases in a particular domain where the naming conventions or other terminology used in the chosen domain does not mean the same in other domain.
3. Using topic maps, one can use analogical reasoning to form hypotheses and, thus, formulate and execute experiments to produce theories.

5.2.4   Advantages of Topic Maps

The main benefits of this approach are:

1. Topic maps help to provide more expressive and detailed metadata and classification systems than the conventional library science approaches.
2. It gives precise searches and navigation systems that are more productive and flexible.
3. It helps in the creation of new knowledge products from the same library resource.
4. Many tools are available for creating and visualizing topic maps.
5. The format used in topic maps is offered interoperability features.

5.2.5 Disadvantages of Topic Maps

The significant disadvantages of topic maps are:

1. The application of topic maps in the practical world is still in its infancy.
2. It is a comparatively new technology, so experts working on topic maps might be harder to find than experts working on traditional library classification schemes in the field of research.
3. As topic maps allow users to define the structure of maps themselves, this leads to the effort of building a classification system, which is a very difficult and lengthy task.

5.3 Network Text Analysis and Libraries

Network analysis methods have been actively used by researchers and library practitioners in the LIS domain over the years to perform content analysis, co-word analysis, and knowledge graphs. It can be applied to numerous resources such as job advertisements, journal articles, open-ended interview transcripts, reference queries, book chapters, books, interviews, essays, newspaper headlines and articles, websites, discussions, conversations, audio recordings, advertising, television and images, speeches, social media data, historical documents, informal conversation, email communication, or any other occurrence of communicative language to visualize the critical nodes (words or documents) and their relationship within the corpus of text.

Nathaniel D. Porter is the Social Science Data Consultant and Data Education Coordinator in the University Libraries and Affiliated Research Faculty in Sociology at Virginia Tech. His research focuses on best practices for teaching with data and transparent/reproducible analysis of online, network and qualitative data.

Story: Visualizing Topic Networks in Personal Correspondence—The Richards/Turner Letters
The Viral Networks Workshop invited scholars of medical history from across the country to answer the question "How can network analysis enhance traditional historical scholarship?" by providing training, collaboration, and consultation applying networks to their own work in preparation of the chapters for the open-access book "Viral Networks." The initiative was a collaboration between Virginia Tech, the NEH, and the NLM. As a Social Science Data Consultant at Virginia Tech University Libraries, I was invited to train and assist with network analysis. In addition to teaching participants

(continued)

how to structure network data and use one visualization tool (Cytoscape), I offered individual consultation for each project. One project, by Katherine Cottle, provided the opportunity to combine networks with NLP. Data were drawn from an online archive of letters between Esther Richards and Abby Turner, two pioneering early-twentieth-century female psychiatrists. Python was used to download the letters, extract text and metadata, and build an LDA topic model with TF-IDF weighting. The co-occurrence network of topics was visualized with Cytoscape, and word clouds of weighted frequencies for high loading in each topic were used, providing a visual "histology reading" (Viral Networks, p. 151) that showed the connections between key concerns in the work and home lives of the correspondents and enriched the close reading approaches used elsewhere in the chapter. The models used can be confusing or intimidating to non-programmers, but the additional processing steps (like visualizing words in topics) can help to communicate findings to general audiences. The chapter, Python script, Cytoscape project, and static and interactive figures are all freely available online in hopes that they will inspire and equip others to analyze topic networks in historical corpora. The book also includes a chapter on getting started with network analysis in the humanities with advice on how to prepare and use data (including output from text mining and NLP) targeted to humanities researchers with limited data or programming experience.
References

1. *Viral Networks Book (Cottle is Ch 6, tutorial is chapter 10)—https://doi. org/10.21061/viral-networks*
2. *Data/viz/code for chapter and book— https://doi.org/10.7294/284t-bf10*
3. *Paper on Data support at VT libraries—Andrea Ogier, Anne Brown, Jonathan Petters, Amr Hilal and Nathaniel Porter. 2018. "Enhancing collaboration across the research ecosystem: Using libraries as hubs for discipline-specific data experts." In PEARC '18: Proceedings of the Practice and Experience on Advanced Research Computing. https://dl. acm.org/doi/10.1145/3219104.3219126*

Manisha Bolina is the Vice President of Business Development at Yewno Inc.

Story: Yewno Discover—A Next-Generation Research Tool
I am sharing information about Yewno Discover, a next-generation research tool that enables library patrons to go beyond keyword searching, by searching through concepts and visualizations. Yewno is a private company and has over 100 customers in 10 different countries around the world. The technology was incubated at Stanford University in California, and today we have academic institutions and public libraries using Yewno Discover. The

(continued)

roots of Yewno's technology go back to 2009 where the founder of Yewno was doing a PhD at King's College London in Econophysics creating an algorithm to read large amounts of structured and unstructured data. The algorithm was then used in a biotech company in Switzerland for drug repurposing for rare diseases.

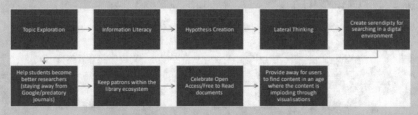

In 2015 our founder was introduced to Prof. Michael Keller, Vice Provost and Library Directory at Stanford, where he was invited to include Yewno's technology. Yewno's patented next-generation AI technology mimics the way the human mind works to read the full text of any document and extract meaning in the form of a concept—this means that it can enable a more granular understanding across a multidisciplinary content set making millions of connections. For the first time, this enables libraries to improve discoverability of content away from metadata, supercharging the library ecosystem. The birth of this technology has led to the creation of Yewno Discover, a next-generation cloud-based AI research tool, enabling better usage and leveraging discoverability of subscribed and "Free to Read" content and understanding concepts through visualizations.

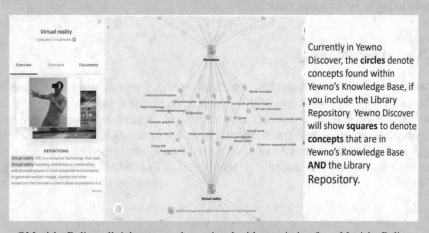

Currently in Yewno Discover, the **circles** denote concepts found within Yewno's Knowledge Base, if you include the Library Repository Yewno Discover will show **squares** to denote **concepts** that are in Yewno's Knowledge Base **AND** the Library Repository.

(continued)

Yewno's AI technology uses a blend of computational linguistics, graph theory, and machine learning. Yewno has created a knowledge graph where it ingests the full text of academic content using a neural network to mimic the human mind extracting concepts and creating an inferential chain of connections. Once the library has implemented it, it is very easy to use—see our YouTube channel for demos[a]. Library budgets can be barrier and those who are resistant or do not understand AI benefits for the library. Our technology is based in Amazon Web Services (AWS). Yewno has a team of data scientists and engineers that keep the maintenance of the tool. Data feeds are provided by feeds sent to Yewno by the publisher so we maintain these on a regular basis.

[a] *https://www.youtube.com/channel/UCMVngx0NmbcPKaYwP1Syutw*

5.3.1   Use Cases

5.3.1.1   Visualizing Ontology

Tanev [18] proposed a new method of using textology (a cluster of co-occurring words) to represent ontologies. Ontologies have major applications (generation of grammar, query expansion, information extraction for information retrieval) in LIS; thus, text networks can bring the paradigm of ontology closer to the text. In addition to textology, topic maps are also semantic-ontological-based methodology organizing library resources by unfolding the structure of knowledge and connecting them with the information resource.

5.3.1.2   Content Analysis

Content analysis is a method that determines the presence and relationship of particular words or concepts within the text. It is often used to determine the occurrence of concepts that are represented by phrases or words in the text (conceptual analysis) and the relationship among concepts in the text (relational analysis). Content analysis is often used with network methods to visualize the concepts and relationship among the words and have been well-researched in LIS.

5.3.1.3   Bibliometrics

Network text analysis has been used widely in the domain of bibliometrics to focus on content. The social network analysis (SNA) methods and techniques were applied in bibliometrics after the emergence of the Internet during the 1990s. Over the years, SNA methods were used to perform co-word analysis. They analyze word/keyword co-occurrence networks using clustering methodologies to identify the prominent words in a corpus of text/bibliographic data (abstract, title, or keywords) and answer questions like "what are the key research sub-domains/topics within a certain domain?" or "how do these sub-domains/topics are related to one other?". It uses the links between documents or words to understand (i) the structure of science, (ii) concepts/themes, and (iii) the association among these concepts.

5.3.1.4   Knowledge Graphs

Knowledge graphs/representations are semantic networks that analyze information using relationships. It has been applied in question-answering system, semantic search, and intelligent service. Using knowledge graph, a library can search keywords based on the entity and relationship, for instance, *Talk to Books* system by Google is a search engine that understands the user's question and content of each book and then matches the information accurately. Knowledge graphs can help:

1. To find resources that may not be available by keyword search as it is based on semantic relation
2. To connect library collection on semantic level, thus improving the efficiency of information acquisition and allowing the users to find different library resources that are connected on semantic level
3. To connect library resources from different institutions or sources
4. To upgrade library's reference service where the QA system responds to users based on semantic analysis of questions and the reasoning ability of the knowledge graph

5.3.1.5   Knowledge Management

Text networks can be used to determine the information flow in knowledge sharing for expert (central node) localization, identification of knowledge communication, and analysis of structure of intra- and inter-organizational knowledge flow in an informal setting [19]. It can help to identify the knowledge sharing barriers or bottlenecks and the group of people with common interest in a particular knowledge area.

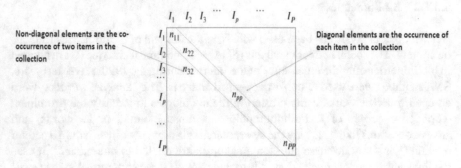

Fig. 5.5   Example of co-occurrence matrix

5.4   Case Study: Network Text Analysis of Documents Using Two Different R Packages

5A: Bibliometrix

Dataset   A plain text file[25] containing bibliometric data of 2254 documents retrieved from Web of Science (WoS) for 2019 for the query on *malaria*.

About the Tool   Refer to Chap. 10, Sect. 10.2.1, to know more about R.

Theory Bibliometrix[26] is an open-source package of R that can be used to launch the biblioshiny app. It is very easy to use even for those who have no coding skills. This package can import bibliometric data from Scopus, Web of Science, PubMed, Dimensions, and Cochrane databases to perform various bibliometric analyses including word co-occurrence network. We used the biblioshiny app to perform word co-occurrence analysis on the retrieved bibliometric data. In word-occurrence network, the starting point is a co-occurrence matrix (Fig. 5.5). The non-diagonal elements of the matrix measure how many times the words appear together in the same corpus (keyword list, title, abstract, etc.) for any two words in the collection. In contrast, the diagonal elements determine the occurrence of each word in the collection. A text network is then prepared using the co-occurrence matrix, where each node represents a word. The node size is proportional to the word occurrence (diagonal elements of the matrix), whereas the edge size is proportional to word co-occurrence (non-diagonal elements of the matrix). We can also detect different communities of words and qualitative measures such as betweenness centrality and PageRank.

---

[25] https://github.com/textmining-infopros/chapter5/blob/master/5a_dataset.txt.
[26] https://bibliometrix.org/.

Methodology The `bibliometrix` library was loaded in `RStudio` to launch the `biblioshiny` app on the host's web browser.

```
#Load libraries
library(bibliometrix)
biblioshiny
```

The following screenshots demonstrate the steps which were taken to perform word-occurrence network:

Step 1: The bibliometric plain text Web of Science file was loaded into the shiny app, followed by saving the PDF of the "main information"[27] of the data (Fig. 5.6).

*Step 2: Word co-occurrence network analysis was performed on 2254 documents' **titles** on malaria using "Jaccard's Index" similarity measure, "Louvain" clustering algorithm, and "Fruchterman & Reingold" network layout for top 50 nodes (Fig. 5.7).*

Step 3: The table containing the values for betweenness, closeness, and PageRank centrality measure for the clusters and words was then saved in CSV format (Fig. 5.8).[28]

Results Figure 5.9 presents the word co-occurrence network for the top 50 words representing the literature indexed in the Web of Science (WoS) database on malaria disease for the year 2019. It was determined from Fig. 5.9 and Table 5.1 that there were three central communities (clusters) which were represented by green, red, and blue colors in the figure. Cluster 1 (red color) consisted of 17 words (nodes) about the parasites *Plasmodium falciparum* and *Plasmodium vivax* that cause malaria disease with an emphasis on the parasite's genetic, molecular, response, and detection rates. The nodes *plasmodium* and *falciparum* had the highest score for all the centrality measures in cluster 1 followed by the nodes *infection*, *blood*, *vivax*, and *human*. Cluster 2 (blue color) consisted of 20 words (nodes) about *malaria* with a focus on its risks, treatment, transmissions, prevalence, clinical trials and factors, Africa, Uganda, and children. In cluster 2, the node *malaria* had the highest value for all the centrality measures followed by *study*, *children*, *associated*, *analysis*, *treatment*, and *transmission*. Cluster 3 (green color) consisted of 13 words (nodes) about *Anopheles* which is the genus of mosquitoes that transmit malaria to humans with an emphasis on gambiae species, vector, evaluation, drug resistance, and activity. The node *anopheles* had the highest value for all the centrality measures in cluster 3 followed by *resistance*, *antimalaria*, *evaluation*, *drug*, *vector*, and *species*.

---

[27] https://github.com/textmining-infopros/chapter5/blob/master/5a_results_main_information.pdf.

[28] https://github.com/textmining-infopros/chapter5/blob/master/5a_results_coword_network_analysis.csv.

(a)

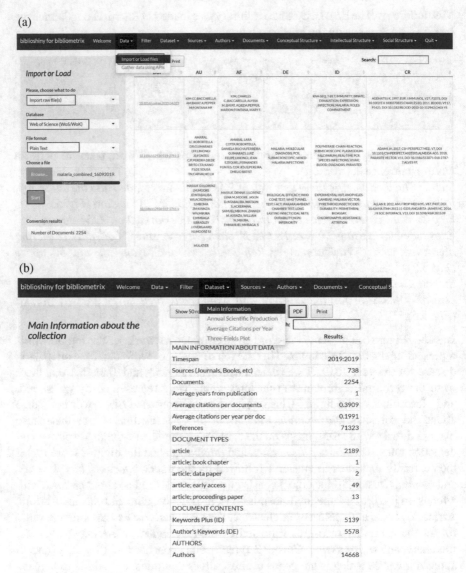

(b)

Fig. 5.6 Screenshots showing step 1: (**a**) Loading of data. (**b**) Determination of main information about the data

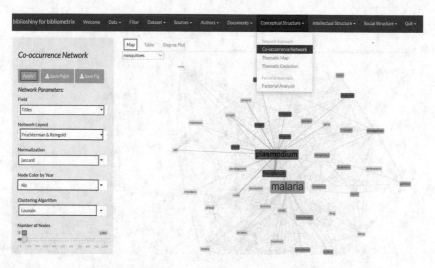

Fig. 5.7 Screenshot showing step 2

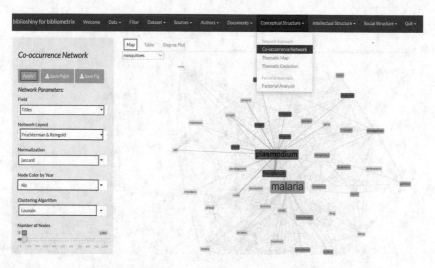

Fig. 5.8 Screenshot showing step 3

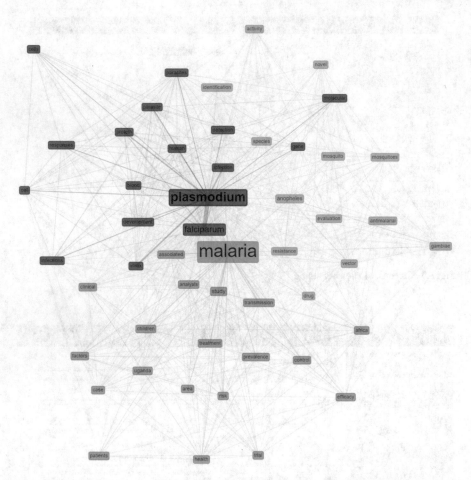

Fig. 5.9 Word co-occurrence network of titles on malaria disease in Web of Science in 2019

5B: Textnets

Prerequisite Familiarity with the R language.

Virtual RStudio Server You can reproduce the analysis in the cloud without having to install any software or downloading the data. The computational environment runs using BinderHub. Use the link (https://mybinder.org/v2/gh/textmining-infopros/chapter5/master?urlpath=rstudio) to open an interactive virtual RStudio environment for hands-on practice. In the virtual environment, open the `text_network_analysis.R` file to perform network text analysis.

Virtual Jupyter Notebook You can reproduce the analysis in the cloud without having to install any software or downloading the data. The computational

Table 5.1 Centrality measure values for top 50 words

Node	Cluster	Betweenness	Closeness	PageRank
Cells	1	0.027294	0.011494	0.007595
Responses	1	0.378969	0.012658	0.012013
Detection	1	1.210888	0.013699	0.012965
Infections	1	0.42973	0.013158	0.011858
Plasmodium	1	137.285	0.02	0.102005
Falciparum	1	50.42198	0.019608	0.067608
Parasites	1	0.86298	0.013333	0.013613
Protein	1	0.912554	0.013514	0.016405
Cell	1	0.611666	0.012821	0.011796
Development	1	0.992917	0.013889	0.011448
Infection	1	5.397887	0.016667	0.026092
Human	1	1.537482	0.014286	0.015808
Blood	1	4.851893	0.015385	0.020881
Vivax	1	1.910235	0.014925	0.02095
Gene	1	0.313928	0.012821	0.010859
Parasite	1	0.880741	0.013699	0.016056
Molecular	1	0.675651	0.012987	0.010908
Malaria	2	309.1364	0.020408	0.127467
Efficacy	2	0.956085	0.012987	0.011908
Control	2	0.484524	0.012987	0.011002
Analysis	2	3.008161	0.016393	0.016435
Clinical	2	0.841131	0.013514	0.012674
Transmission	2	1.630175	0.014706	0.018369
Children	2	4.457946	0.014706	0.02541
Risk	2	1.363823	0.014085	0.017669
Factors	2	0.629126	0.012987	0.017627
Patients	2	0.088771	0.011494	0.009413
Africa	2	0.626614	0.012987	0.010307
Study	2	13.5377	0.017241	0.034577
Area	2	0.466493	0.012987	0.011798
Uganda	2	0.772082	0.013333	0.011358
Trial	2	0.27268	0.0125	0.009319
Health	2	1.305334	0.012987	0.014451
Treatment	2	2.21992	0.014085	0.018425
Associated	2	3.550089	0.014925	0.021009
Prevalence	2	0.727069	0.013514	0.013526
Case	2	0.465879	0.012821	0.011519
Vector	3	1.266642	0.013333	0.016578
Mosquitoes	3	0.79737	0.013158	0.011555
Identification	3	0.80872	0.013333	0.011332
Anopheles	3	14.14053	0.015152	0.032911

(continued)

Table 5.1   (continued)

Node	Cluster	Betweenness	Closeness	PageRank
Species	3	1.11182	0.013699	0.012487
Evaluation	3	1.723466	0.013333	0.012031
Gambiae	3	0.166766	0.011905	0.010848
Novel	3	0.422853	0.0125	0.008988
Antimalarial	3	1.897918	0.012658	0.012765
Drug	3	1.378278	0.013514	0.013159
Mosquito	3	0.776997	0.013333	0.011407
Resistance	3	7.876115	0.016129	0.023999
Activity	3	0.390756	0.011905	0.008816

environment runs using BinderHub. Use the link (https://mybinder.org/v2/gh/textmining-infopros/chapter5/master?filepath=Case_Study_5B.ipynb) to open an interactive virtual Jupyter Notebook for hands-on practice.

Dataset   A CSV file[29,30] containing metadata for 98 electronic theses and dissertations (ETDs) from Shodhganga database from 2013 to 2017 in library and information science (LIS) subject.

About the Tool   Refer to Chap. 10, Sect. 10.2.1, to know more about R.

Theory   Textnets[31] is an R package that conducts automated text analysis to detect the latent themes in unstructured textual data using network techniques. It provides an alternate technique to topic models to synthesize unstructured textual data using natural language processing and graph theory/network analysis.

Methodology and Results   The libraries and the dataset required to perform network text analysis in R were loaded.

```
#Load libraries
library(textnets)

#Load dataset data <- read.csv("https://raw.githubusercontent.com/
textmining-infopros/chapter5/master/5b_dataset.csv?
token=ARBWLQ7FRPMWAXI27I6OGOTACZLIW")
```

The titles of the ETDs were then pre-processed to remove stopwords, punctuation, number, and whitespace and were prepared for network analysis using all parts of speech but were limited to noun compounds and nouns. "It is preferable to

---

[29] https://github.com/textmining-infopros/chapter5/blob/master/5b_dataset.csv.

[30] Note: There is some glitch in the package that it recognizes the text column only if named as *textvar*. Therefore, we replaced the *title* column with *textvar*.

[31] https://github.com/cbail/textnets.

create networks based only on nouns and noun compounds as previous studies [20] have shown that they are useful in mapping the topical content of a text than other parts-of-speech such as verbs or adjectives" [9].

```
#Preparing Text with Nouns
prep<- PrepText(data,
groupvar = "Researcher.Name", textvar = "textvar",
node_type = "words", tokenizer = "words", pos = "nouns",
remove_stop_words = TRUE, remove_numbers = TRUE,
compound_nouns = TRUE)
```

Using the pre-processed data, we created the network graph that "outputs an igraph object based on a weighted adjacency matrix where the rows and columns correspond to words. The cells of the matrix are the transposed cross-products of the term-frequency inverse-document frequency (TF-IDF) for overlapping terms between the words and the matrix product of TF-IDF cross-product" [9].

```
#Creating Text Network
network <- CreateTextnet(prep)
```

We then visualized the network by creating "a network diagram whose nodes are colored by their cluster or modularity class" [9]. It employs the *network backbone* technique by using a tuning parameter called *alpha*, "which deletes edges using a disparity filter algorithm to trim edges that are not informative" [9].

```
#Visualizing the Text Network
    VisTextNet(network, alpha = .15, label_degree_cut = 0,
betweenness = FALSE)
    print(VisTextNet(network, alpha = .15, label_degree_cut = 0,
betweenness = FALSE))
```

The words were then grouped "according to their similarity in order to identify their latent theme across texts by clustering the words within the text network by applying the Louvain community detection algorithm, which uses the edge weights and determines the number of clusters within the network. The communities function outputs a dataframe with the cluster or modularity class to which each word is assigned" [9] and was saved in CSV format.[32] Figure 5.10 and Table 5.2 represent the 22 clusters/communities for the text network with 232 words (nodes).

---

[32] https://github.com/textmining-infopros/chapter5/blob/master/5b_results_communities.csv.

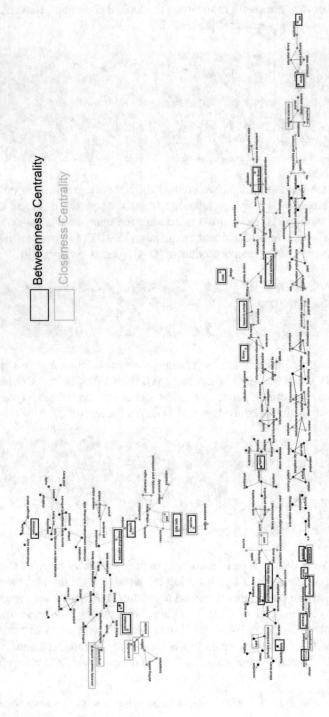

Fig. 5.10 Text network showing 22 communities of latent topics

Table 5.2 Identified communities/clusters using Louvain Community Detection Algorithm

Cluster 1	Cluster 2	Cluster 3	Cluster 4	Cluster 5
Access	Dictionary	Collection development policy	E content	Area
Article	Sanskrit language	Graduate college library	Hyderabad karnataka region	Information literacy
Bibliometric		Karnataka university dharwad	Knowledge	Management student
Citation		Procedure	Wisdom	
Journals				
Recency pattern				
Research				
Science				
Webometric analysis				

Cluster 6	Cluster 7	Cluster 8	Cluster 9	Cluster 10
Government	Analysis	Culturalcontribution	Colleges	Bangalore city
Jharkhand	Ebsco database	Performance evaluation	Industry library	Chandigarh
Library information centre	Evaluation	South region Ahmednagar district	Information product source	Citation analysis
User behaviour	Finance		Information science education	Delhi
	Librarian perspective		Job market	Engineering
	Literature		Librarian	Faculty member
	Region		Library	Haryana
	Science blogs		Maharashtra	Institution
	Website		Marketing	Journal
	Westbengal		Nba	Material sciences
			Relevance	Punjab
			Technology adoption	Study
			User study	User awareness

(continued)

Table 5.2 (continued)

Cluster 18	Cluster 19	Cluster 20	Cluster 21	Cluster 22
Behaviour	Effectiveness	Accreditation	Architecture college library	Generation
Case study	Gujarat state	Administration	Art	Importance
Center	Information service	Communication technology	Goa	India
Health information	Researcher	Congregations	India radio	Property information
Information need	Universities researcher	Drdo library	Information source	Role
Investigation	Visvesvaraya	Education programme	Kanyakumari district	University grant commission
Preception awareness		Functioning	Punecity	Vidharbha region
knowledge information need				
Public		Impact	Quality assessment	
Source		Maduraikamaraj university area	Resource	
Tourist		Mumbai	Science college library	
Tourist information system		Seminary	U.p	
		Utilisation	Uttarakhand	

Cluster 11	Cluster 12	Cluster 13	Cluster 15
Acceptance	Assam	Department	Btisnet library
Activity	Challenge	Discipline	Consortium
Attitude	Child library	Information communication technology skills	Feasibility study
Capital region	Growth	Karnataka	Habit
Classification scheme	Job analysis	Maharashtra state	Indion
Collection development	Prospect	Maharashtra state	Institute
College library	Reality	Researchers	Software professional
E resource	Research output	Resource development	
Ict environment	Source library management software	Sciences	
Information technology	Technology institute	Universities	
Jalgaon district be		Web information resource	
Jurisdiction			
Library environment			
North Maharashtra			
Professional			
Pune			
Solapur university			
Student teacher			
University			
Use			
User			

(continued)

Table 5.2 (continued)

Cluster 14	Cluster 16	Cluster 17
Applications	Andhra pradesh	Application
Automation	Branch	Ayurveda
Bangalore	Collection management	Climate
Centre	College	Comparison
Child	Dakshina kannada	Context
City	District	Dibrugarh district
Design	Engineering college library	Dissemination
Development	Faculty	Effectiveness study
Education library	Iit	Information
Jammu division	Information resource	Journal collection management transition trend
Literacy	Karnataka state	Kerala
Maharashtra government	Karnataka state law university hubli	Medicine
Management college library	Law library	Organization administration
Model	Literacy skills	Pattern
Network	Networking	Plant
Network architecture	Polytechnic	Preparation
Prototype model	Programme	Productivity
Quality management	Review	Professional commitment
Research scholar	Skills	Reference
Security	Suburaban	Regard
Source software	University research scholar	Science professional
System		Scientist

```
#Analyzing Text Networks
communities <- TextCommunities(network)
write.csv(communities, 'communities.csv')
```

Lastly, we calculated the measure of influence or centrality for each node and saved it in CSV format.[33] The top 25 nodes with high betweenness and closeness centrality were marked with black and yellow boxes, respectively (Fig. 5.10). It was observed that many of the top nodes with high betweenness measure had high closeness measure too. The nodes with high betweenness centrality "play an important role in the transmission of data in the network" [21], thus considered as the topics "with an interdisciplinary approach in the field" [21] of LIS, whereas the nodes with "high closeness centrality are more effective and central in the network and have greater accessibility than other nodes" [21].

```
text_centrality <- TextCentrality(network)
write.csv(text_centrality, 'text_centrality.csv')
```

References

1. Diesner, J, Carley, KM (2005) Revealing social structure from texts: Meta-matrix text analysis as a novel method for network text analysis. In: Narayanan VK, Armstrong DJ (eds) Causal mapping for research in information technology, Harrisburg, pp 81–108
2. Hunter S (2014) A novel method of network text analysis. Open J Modern Linguist 04:350–366. https://doi.org/10.4236/ojml.2014.42028
3. Czachesz I (2016) Network analysis of biblical texts. J Cogn Historiography 3:43–67. https://doi.org/10.1558/jch.31682
4. Lemaire B, Denhiere G (2004) Incremental construction of an associative network from a corpus. In: Proceedings of the annual meeting of the cognitive science society. https://escholarship.org/uc/item/3k98b25s. Accessed 26 Nov 2020
5. Serrano JI, Iglesias A, Castillo MD del (2007) Modeling human reading in conceptual networks for text representation and comparison. In: 2007 international joint conference on neural networks, Orlando, USA, pp 613–618. https://doi.org/10.1109/IJCNN.2007.4371027
6. Palshikar GK (2007) Keyword extraction from a single document using centrality measures. In: Ghosh A, De RK, Pal SK (eds) Pattern recognition and machine intelligence. PReMI 2007. Lecture notes in computer science, vol 4815. Springer, Berlin, Heidelberg
7. Bail CA (2016) Combining natural language processing and network analysis to examine how advocacy organizations stimulate conversation on social media. Proc Natl Acad Sci 113:11823–11828. https://doi.org/10.1073/pnas.1607151113

---

[33] https://github.com/textmining-infopros/chapter5/blob/master/5b_results_centrality.csv.

8. Opsahl T (2013) Triadic closure in two-mode networks: Redefining the global and local clustering coefficients. Social Networks 35:159–167. https://doi.org/10.1016/j.socnet.2011.07. 001

9. Bail C (2020) Text networks. https://sicss.io/2020/materials/day3-text-analysis/text-networks/ rmarkdown/Text_Networks.html. Accessed 19 May 2021

10. Disney A (2010) Social network analysis 101: centrality measures explained. https:// cambridge-intelligence.com/keylines-faqs-social-network-analysis/. Accessed 13 May 2021

11. Ianni M, Masciari E, Sperlí G (2020) A survey of Big Data dimensions vs Social Networks analysis. J Intell Inf Syst. https://doi.org/10.1007/s10844-020-00629-2

12. Paranyushkin D (2011) Identifying the pathways for meaning circulation using text network analysis. Venture Fict Pract 2(4). https://noduslabs.com/wp-content/uploads/2012/04/ Pathways-Meaning-Text-Network-Analysis.pdf. Accessed 19 May 2021

13. Paranyushkin D (2010) Text network analysis. https://issuu.com/deemeetree/docs/text-network-analysis. Accessed 27 Nov 2020.

14. Hunter S (2014) A novel method of network text analysis. Open J Modern Linguist 4(2):350–366. https://doi.org/10.4236/ojml.2014.42028

15. Pepper S (2002) The TAO of topic maps: Finding the way in the age of Infoglut. Ontopia. https://ontopia.net/topicmaps/materials/tao.html. Accessed 27 Nov 2020

16. Wlodarczyk B (2012) Topic Map Library = Better Library: an Introduction to the "National Library of Poland" Project. In: World library and information congress: 78th IFLA general conference and assembly, Helsinki. https://www.ifla.org/past-wlic/2012/117-wlodarczyk-en. pdf

17. Garshol LM (2004) Metadata? Thesauri? Taxonomies? Topic Maps! Making sense of it all. J Inf Sci 30(4):378–391. https://doi.org/10.1177/0165551504045856

18. Tanev H (2014) Learning textologies: Networks of Linked word clusters. In: Biemann C, Mehler A (eds) Text mining. Theory and applications of natural language processing. Springer, Cham. https://doi-org.lib-ezproxy.concordia.ca/10.1007/978-3-319-12655-5_2

19. Helms R, Ignacio R, Brinkkemper S, Zonneveld A (2010) Limitations of network analysis for studying efficiency and effectiveness of knowledge sharing. Electron J Knowl Manag 8(1):53–68

20. Rule A, Cointet J-P, Bearman PS (2015) Lexical shifts, substantive changes, and continuity in State of the Union discourse, 1790–2014. PNAS 112:10837–10844. https://doi.org/10.1073/ pnas.1512221112

21. Sedighi M (2016) Application of word co-occurrence analysis method in mapping of the scientific fields (case study: the field of Informetrics). Library Review 65:52–64. https://doi. org/10.1108/LR-07-2015-0075

Chapter 6
Burst Detection

Abstract This chapter provides a theoretical framework for burst detection, including its advantages, disadvantages, and other essential features. It further enumerates various open-source tools that can be used to conduct burst detection and discusses the use cases on how the information professionals can apply it in their daily lives. The chapter is followed by a case study using two different tools to demonstrate the application of burst detection in libraries.

6.1 What Is Burst Detection?

A time series can be defined as a sequence of events or observations over time. A time-series data can be discrete or continuous, for instance, Wikipedia editing activity or Google Trends[1] showing results for Google searches for different queries. Time-series events are commonly represented by lists where a record represents each event with temporal attribute(s) (Fig. 6.1). The following are some of the common patterns in a time-series data:

1. Trends: It corresponds to the general tendency of time-series data to be increasing, decreasing, having stability, or be in a cyclic tendency.
2. Seasonality: These are the repetitive and predictable movement around a trend line such as cyclic variations of flu infections or harvesting of crops.
3. Bursts: These correspond to the identification of sudden spurt of activity, sometimes in response to an external event such as a disease.

Burst detection is a temporal analysis that "aims to identify the nature of phenomena represented by a sequence of observations such as patterns, trends, seasonality, outliers, and bursts of activity" [1]. It can be used to (i) understand the temporal distance such as the most emerging or trending terms, growth of terms, and latency/peak or (ii) forecasting, that is, predicting, future values of the time-series variables. Various algorithms [2–6] perform burst detection, but Kleinberg's

---

[1] https://trends.google.com/.

(a)

Subject header	Date	Time
Meeting for project	1/1/2021	10:03 AM
Invitation to give talk	1/1/2021	10:05 AM
Review paper	1/1/2021	12:01 PM
Happy New Year!	1/1/2021	12:01 PM
To do list reminder	1/1/2021	12:04 PM

(b)

Month	Day	Year	Time	Rank
12	09	2021	20:00	75,000
12	09	2021	22:00	65,975
12	09	2021	23:00	70,313
12	10	2021	2:00	98,209
12	10	2021	3:00	106,957

Fig. 6.1 Examples showing temporal data. (**a**) Emails. (**b**) Amazon books' rankings

algorithm [7] is the most popular one that detects bursts in discrete batches of events. This algorithm identifies the bursts of topics for a temporal stream of documents based on the Hidden Markov Model. "Originally developed to detect topics in email chains, Kleinberg's method assumes that terms in documents are emitted by a two-state automaton. The automaton may spontaneously transit from a non-bursty state to a bursty state, or vice versa" [8]. It is "commonly applied to identify words that have experienced a sudden change and increase in the frequency of occurrence. The algorithm generates a list of word bursts in the document stream, ranked according to the burst weight, together with the intervals of time in which these bursts occurred. The burst weight depicts the intensity of the burst, i.e., how great the change in the word frequency that triggered the burst" [1]. Thus, burst detection helps in the classification and detection of keywords and in "maintaining a record of the event based on related features" [5]. These features help to find the trending pattern of events meticulously at different times.

6.1.1   How to Detect a Burst?

Textual data can be imagined as a discrete time series where a sequence of observations/events occurs over time. These observations happen at regularly spaced intervals such as every month, week, volume, issue, or year. *Bursts* are the large number of events taking place within a defined period. In contrast, *burstiness* is defined as the events concentrated within a short period of intense activity followed by long intervals of inactivity. An event can be planned or unplanned and can be trending or non-trending, which is directly associated with the event's frequency. For instance, planned events such as IPL 2021 Cricket is an example of bursty event, whereas American Idol Premier is also a planned event but non-bursty. "Similarly, unplanned events such as Cyclone 2021, India was bursty, and a minor road accident is non-bursty. The bursty behavior is directly associated with the diffusion of information over social media or a network" [5]. There are two approaches to detect a burst:

1. **Analyzing complete data stream**: This approach is not appropriate for identifying bursts for every occurrence of an event.
2. **Using multiple time windows**: This approach can identify bursts in real time by decreasing window size and calculates the combined occurrence of events within the time window. Moreover, this approach helps in storing the collection of every event for each period.

Real-time detection of a burst at an early stage in emails, news, blogs, research articles, tweets, comments, and Internet bulletin boards is becoming gradually significant in several fields where topics appear and grow in intensity rapidly. To detect a burst and report it as early as possible is quite valuable for essential decision-making processes within a discipline. Various methods, like content analysis [9], clustering [10–16], search [17], and personalization [18], have been proposed for burst detection.

6.1.2 Comparison of Burst Detection with Others

Topic Detection and Tracking (TDT)[2] was first used to solve the issue of tracking topics for time-ordered problems, but with the advancement in computer science research, Latent Dirichlet Allocation (LDA)[3] (see Chap. 4) became the most popular approach for topic modeling. However, topic modeling has many disadvantages, for instance, "the lack of interpretability of the results and the difficulty in coherently linking LDA topics together between subsequent time-steps" [8]. In contrast to LDA, burst detections "first identify the bursty terms in a dataset, and then cluster them together into topics" [8]. Singh and Kumari [5] compared their burst detection technique (*BUrST*) with other state-of-the-art methods like LDA, graph-based feature-pivot topic discovery, document-pivot topic detection, and BN-grams method. They found *BUrST* to be proficient in detecting valuable patterns of interest. Figure 6.2 shows the result for a timestamped text for which the Kleinberg's algorithm identifies words that burst. In Fig. 6.2, x does not burst as it is present in all the years, whereas y bursts more than z. The algorithm results in a *level of burstiness, weight of the burst, length of the burst*, and *start and end year* for a

---

[2] https://ciir.cs.umass.edu/tdt.

[3] https://dl.acm.org/doi/10.5555/944919.944937.

Year	Words
1996	x
1997	x y
1998	x y
1999	x
2000	x
2001	x y y
2002	x y y
2003	x y y
2004	x y y
2005	x y
2006	x y
2007	x y
2008	x
2009	x
2010	x
2011	x
2012	x
2013	x z
2014	x z
2015	x z
2016	x z
2017	x z
2018	x z
2019	x
2020	x
2021	x

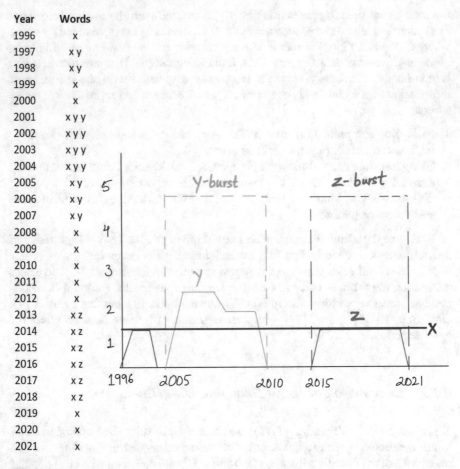

Fig. 6.2 Example showing bursts

given stream of events where every event is a set of keywords with a timestamp. Thus, burst detection identifies time intervals with an unusually high frequency of a specific keyword.

6.1.3 How to Perform Burst Detection?

For burst detection, texts need to be cleaned effectively to get a profound representation of the data. The typical steps to perform burst detection are:

1. Pre-processing of text. It includes (i) removal of replies, mentions, URLs, hashtags, and retweets, (ii) correction of spelling errors using a specific dictionary such as Hunspell dictionary, (iii) replacing of abbreviations and shorthand

Fig. 6.3 Types of time
slicing

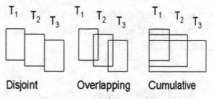

Disjoint   Overlapping   Cumulative

Disjoint: Every row in the original table
is in exactly one time slice

Overlapping: Selected rows are in
multiple time slices

Cumulative: Every row in a time slice is
in all later time slices

notions using a specific dictionary such as SMS (Short Message Service)
dictionary for social media data in addition to other pre-processing tasks covered
in Sect. 3.3.

2. Filtering of particular time slice by seconds, milliseconds, days, hours, minutes,
   fortnights, months, weeks, quarters, decades, years, or centuries for disjoint,
   overlapping, or cumulative time slice (Fig. 6.3); removal of large spikes in the
   data; normalization of data by de-duplication, unit conversion, and adjusting
   time zones; integration and interlinking of different data sources; and finally,
   the aggregation/classification of the data.
3. Performing burst detection.
4. Visualizing the bursts.

6.1.4   Available Tools and Packages

The Sci2 Tool[4] and CiteSpace[5] are the two open-source tools/applications
that non-programmers can utilize to perform burst detection. Programmers can
use Bibliometrix/Biblioshiny[6] and bursts[7] packages in R, and

---

[4] https://github.com/CIShell/sci2/releases/tag/v1.3.0.

[5] http://cluster.cis.drexel.edu/~cchen/citespace/.

[6] https://www.bibliometrix.org/Biblioshiny.html.

[7] https://cran.r-project.org/web/packages/bursts/index.html.

`Metaknowledge`,[8] `burst_detection`,[9] and `pybursts`[10] packages in `Python` to perform burst detection. Most of the tools mentioned above are covered in detail in Chap. 10.

6.1.5   Applications

Burst detection is commonly used to identify activity spikes in email, Twitter, Flickr, Facebook, or news data streams. These bursts of activities can be correlated with external events such as deadlines. It has been applied to several types of documents such as blogs [19, 20], tweets [21–23], spam detection [24], news streams [25, 26], and research articles [8, 27–29]. "Scientific papers tend to enter the world in batches, such as when a new edition of a journal or the proceedings of a conference is published. This violates Kleinberg's underlying assumption that new items enter the dataset in a continuous fashion. It also forces us to impose longer time steps, such as years rather than seconds" [8]. In contrast, several open-access repositories for scientific literature such as PubMed, arXiv, Semantic Scholar, and DBLP contain articles with small intervals compared to the size of the time steps, making them a good data source for burst detection. Bursts can be detected for journal names, author names, keywords, references, terms, or country names used in the abstract or title of a document.

6.1.6   Advantages

1. Scalable.
2. "It can be applied to datasets for which one might have little domain knowledge" [8].
3. It can help "to create a snapshot of the history of a field" [8].
4. It can be helpful "to funding agencies and researchers exploring the research landscape" [8].

6.1.7   Limitations

1. Bursts might pick up trends in language use and style rather than the content of the documents.

---

[8] https://pypi.org/project/metaknowledge/.

[9] https://pypi.org/project/burst_detection/#:~:text=Burst%20detection%20identifies%20time %20periods,submissions%20to%20an%20annual%20conference).

[10] https://pypi.org/project/pybursts/.

2. Detecting burst events is rather difficult because of noisy and sparse social media texts.
3. Bots or fake accounts spreading misinformation/disinformation by re-posting tweets/news/posts make the information containing them bursty on social media.
4. It might also pick up the changes in the construction of the text, for instance, patents vs. papers.
5. It "requires a span of historical data to detect bursts which means that it cannot effectively detect bursts in the earliest years of a dataset" [8].

6.2   Burst Detection and Libraries

In libraries, burst detection can help to track the trending topics discussed by libraries on Twitter or the subject of popular books being borrowed by library users. Library professionals may use burst detection in bibliometric studies to understand the emergence of a concept or apply it to textual data retrieved from social platforms to determine the burst of a concept/topic as a query over a defined epoch and repackage the information in various formats like a state-of-the-art report to cater to the needs of its users.

6.2.1   Use Cases

6.2.1.1   Personalization

Burst detection can provide tailored content to the patrons matching their preferences and habits based on their profiles, that is, providing them with selective dissemination of information (SDI) service.

6.2.1.2   Information Retrieval

Burst detection can enhance the searching process of large documents in the library to aid in the indexing and ranking process of the library's OPAC.

6.2.1.3   Bibliometrics

Burst detection is a prevalent methodology in the domain of bibliometrics. It has been used to study the fastest rising or hot topics or *bursts* in the scientific literature to provide insight into how they evolved. There are many applications of finding bursty terms in the scientific literature, including (i) "early detection might allow funding agencies and publishers to take note of the most promising new ideas and channel new support that way" [8], (ii) "automatically listing the hottest topics over time would give an instant snapshot of the evolution of the field" [8], and (iii) "compiling a corpus of historical bursty terms over time might make it possible to characterize the life cycles that new ideas go through as they develop" [8].

6.2.1.4 Altmetrics

Burst detection can help the libraries to determine the real-time detection of concepts or keywords appearing in the altmetric data related to the research conducted by their faculty.

6.2.2 Marketing

Burst detection can identify the trending hashtags or topics on the library's social media platforms and help the librarian and decision-making authorities to change/improve library products and services.

6.2.3 Reference Desk Service

Burst detection can identify the most bursty words in a corpus of reference questions asked by patrons in person or virtually and can help the librarian prepare in advance for related queries.

6.3 Case Study: Burst Detection of Documents Using Two Different Tools

6A: Sci2

Problem If you have text documents such as research articles, tweets, newspaper articles, electronic theses and dissertations, blog posts, Facebook posts, or reviews and want to identify the emerging/hot/trending topics in the documents over some time.

Goal To identify the trending/emerging terms, subjects, or topics for a particular period.

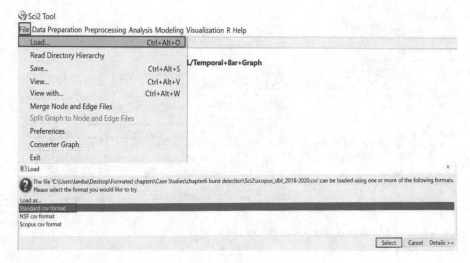

Fig. 6.4   Loading of data in the Sci2 Tool

Dataset   A query `"FUND-ALL (department AND of AND biotechnology AND dbt ANDindia) AND (LIMIT-TO (FUND-SPONSOR,"Department of Biotechnology"))AND (LIMIT-TO (PUBYEAR, 2020) OR LIMIT-TO (PUBYEAR, 2019) OR LIMIT-TO (PUBYEAR, 2018) OR LIMIT-TO (PUBYEAR, 2017) OR LIMIT-TO (PUBYEAR, 2016)) AND (LIMIT-TO (DOCTYPE, "ar")) AND (LIMIT-TO(AFFILCOUNTRY, "India"))"` was searched in the Scopus[11] database for the research articles published with the help of funding received from the *Department of Biotechnology* (an Indian funding agency) in the field of science and technology from 2016 to 2020. A CSV file[12] containing the metadata for 2062 research articles was retrieved from 2018 to 2020.

About the Tool   Refer to Chap. 10, Sect. 10.2.7, to know more about `Sci2 Tool`.

Methodology   The following screenshots demonstrate the steps which were taken to perform burst detection:

Step 1: The data was loaded into the Sci2 Tool (Fig. 6.4).
Step 2: Text pre-processing was performed on the abstracts (AB) of the retrieved research articles (Fig. 6.5).
Step 3: Burst detection was performed on the pre-processed abstracts (Fig. 6.6).
Step 4: The CSV file consisting of the result of burst detection analysis was saved and edited. In the empty column of "End" year, the last year (i.e., 2020 in this case)

---

[11] https://www.scopus.com/home.uri.

[12] https://github.com/textmining-infopros/chapter6/blob/master/6a_dataset.csv.

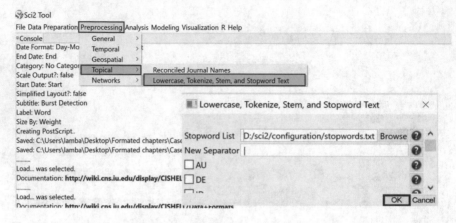

Fig. 6.5 Pre-processing of abstracts in the Sci2 Tool

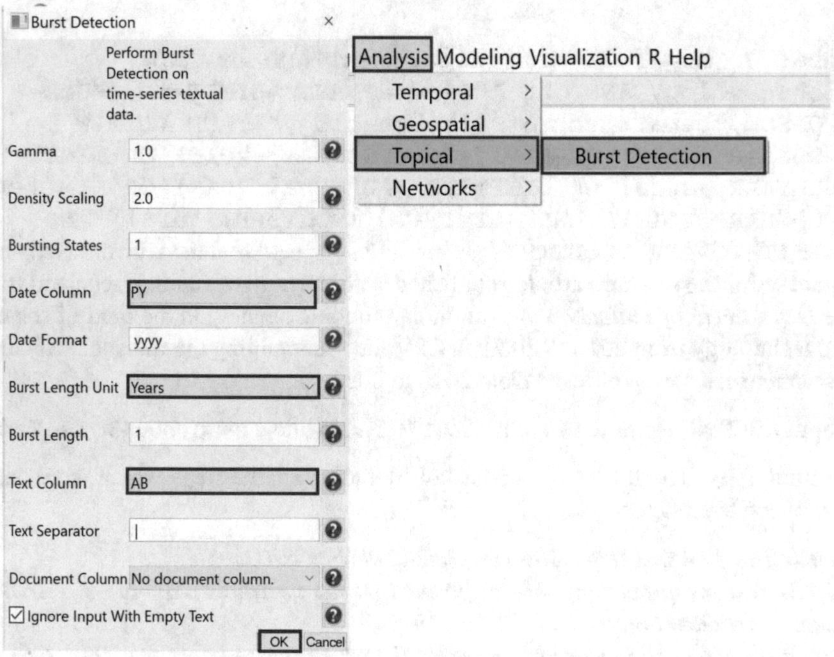

Fig. 6.6 Performed burst detection on abstracts in the Sci2 Tool

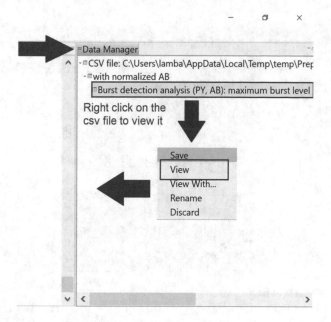

Fig. 6.7 Saving burst detection results

of the analysis was added, and some of the spurious words rows like just, 2019, and 2020 were deleted from the file (Fig. 6.7).
Step 5: The saved CSV file[13] from Step 4 was then loaded in the Sci2 Tool to visualize the bursts (Fig. 6.8).
Step 6: The postscript[14] and bar size[15] files were then saved. The postscript file was visualized using psview[16] tool (Fig. 6.9).

Results Figure 6.10 represents each record as horizontal bars with a specific start and end year, where the x-axis is time and the y-axis is amount. The sliced time table[17] shows two levels of bursts: *level 1* shows data from its own time interval, whereas *level 2* shows the growth from one year to another. The top words with the highest weight and area for each year/period and their associated topic/themes are summarized in Table 6.1. As shown in the table, the popular topic of research by Indian researchers in biotechnology in 2018 was on *histones*; in 2019, it was on *breast cancer*; in 2020, it was on *coronavirus*.

---

[13] https://github.com/textmining-infopros/chapter6/blob/master/6a_maximum_burst_level.csv.

[14] https://github.com/textmining-infopros/chapter6/blob/master/6a_results_horizontal-line-graph.ps.

[15] https://github.com/textmining-infopros/chapter6/blob/master/6a_results_barSizes.csv.

[16] http://psview.sourceforge.net/download.html.

[17] https://github.com/textmining-infopros/chapter6/blob/master/6a_maximum_burst_level.csv.

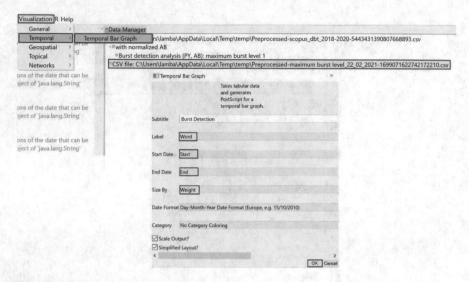

Fig. 6.8 Visualization of bursts

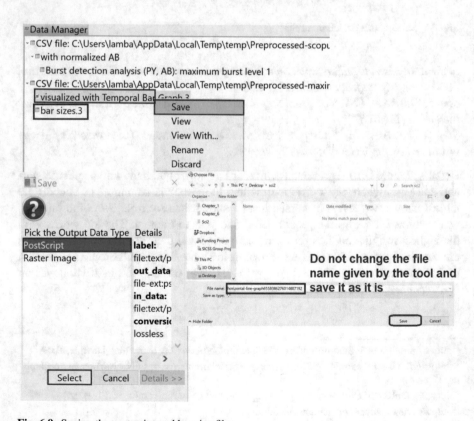

Fig. 6.9 Saving the postscript and bar size files

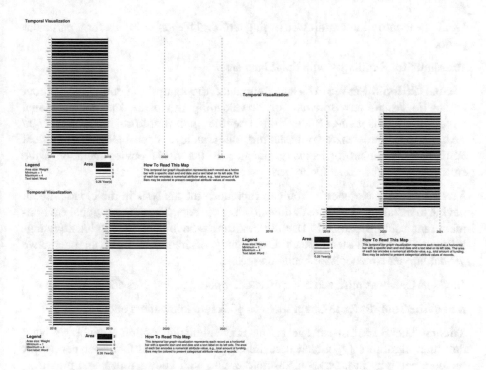

Fig. 6.10 Horizontal line graph

Table 6.1 Summarization of burst results

Level	Period/year	Top 5 words	Area	Topic
1	2018	Label, union, histon, event, hypothes	3.95, 3.93, 3.68, 2.84, 2.78	Histones
2	2019–2020	Threaten, breast, cytometri, machineri, elucid	3.26, 2.60, 2.10, 1.99, 1.95	Breast cancer
1	2020	Covid-19, cov-2, pandem, coronavirus, respiratori	4.23, 4.23, 3.96, 3.43	Coronavirus (COVID-19) pandemic

6B: R

Problem If you have text documents such as research articles, tweets, newspaper articles, electronic theses and dissertations, blog posts, Facebook posts, or reviews and want to identify the emerging/hot/trending topics in the documents over some time.

Goal To identify the trending/emerging terms, subjects, or topics for a particular period.

Prerequisite Familiarity with the R language.

Virtual RStudio Server You can reproduce the analysis in the cloud without having to install any software or downloading the data. The computational environment runs using BinderHub. Use the link (https://mybinder.org/v2/gh/ textmining-infopros/chapter6/master?urlpath=rstudio) to open an interactive virtual RStudio environment for hands-on practice. In the virtual environment, open the `burst_detection.R` file to perform burst detection.

Virtual Jupyter Notebook You can reproduce the analysis in the cloud without having to install any software or downloading the data. The computational environment runs using BinderHub. Use the link (https://mybinder.org/v2/gh/textmining-infopros/chapter6/master?filepath=Case_Study_6B.ipynb) to open an interactive virtual Jupyter Notebook for hands-on practice.

Dataset Data was retrieved using an `aRxiv` package that uses arXiv API[18] in R.

About the Tool Refer to Chap. 10, Sect. 10.2.1, to know more about R.

Theory The `bursts` package is "an implementation of Jon Kleinberg's burst detection algorithm [7] which uses infinite Markov model to detect periods of increased activity in a series of discrete events with known times, and provides" [30] visualizations for the results.

Methodology and Results The `aRxiv` library was used to search 1000 abstracts that contained the query "burst detection." 233 articles[19] were identified for the query from 1993 to 2021.

```
#Load libraries
library(aRxiv)
library(bursts)

  data <- arxiv_search('abs:"burst detection"', limit=1000)
write.csv(data, "6b_dataset.csv")
```

`Kleinberg's` function was used to perform burst detection. It resulted in a dataframe that contained a list of all the "bursts" and analyzed the dates when the documents were submitted in the arXiv repository.[20]

---

[18] https://arxiv.org/help/api/.

[19] https://github.com/textmining-infopros/chapter6/blob/master/6b_dataset.csv.

[20] https://arxiv.org/.

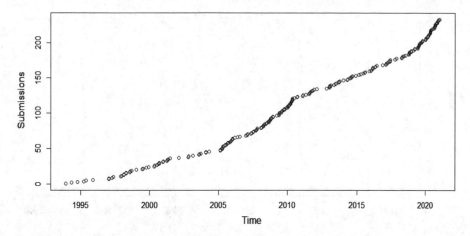

Fig. 6.11 Accumulation of "burst detection" submission in arXiv repository

```
bursts <- kleinberg(as.POSIXct(data$submitted))
```

Figure 6.11 shows the accumulation of "burst detection" submissions in the arXiv repository. It can be observed that submissions in the field of "burst detection" in the arXiv repository are from 1993, and the number of submissions increased in intensity over the years.

```
#Accumulation of submissions
plot(as.POSIXct(data$submitted),
1:length(as.POSIXct(data$submitted)),
xlab='Time',
ylab='Submissions')
```

The bursts were then plotted to present the hierarchical burst structure (Fig. 6.12). Table 6.2 shows (i) the level, (ii) the start date, and (iii) the end date of the identified bursts. Level 1 represents the first event from its first instance (1993) in the arXiv repository to the last (2021). Level 2 represents the additional levels of bursts (2008–2010 and 2018–2021) for the pre-prints in the arXiv repository.

```
#Bursts in submissions
plot(bursts, xaxt = 'n')
axis.POSIXct(1, bursts$start)
```

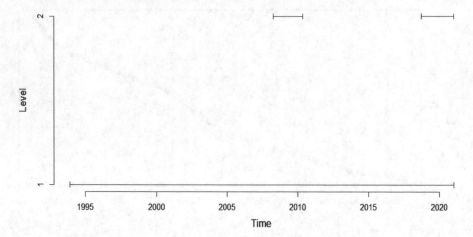

Fig. 6.12 Bursts of "burst detection" in arXiv repository

Table 6.2 Levels of bursts

Level	Start	End
1	11-18-1993	01-12-2021
2	04-07-2008	05-12-2010
2	10-01-2018	01-12-2021

References

1. Lind S (2016) Science of science (Sci2) tool manual. https://wiki.cns.iu.edu/pages/viewpage.action?pageId=1245860#id-4.6TemporalAnalysis(When)-4.6.1BurstDetection. Accessed 22 Feb 2021
2. Zhang X, Shasha D (2006) Better burst detection. In: Proceedings of the 22nd international conference on data engineering. IEEE Computer Society, Washington, DC, pp 146–149
3. Zhu Y, Shasha D (2003) Efficient elastic burst detection in data streams. In: Proceedings of the ninth ACM SIGKDD international conference on knowledge discovery and data mining. ACM, New York, pp 336–345
4. Ryan D (ed) (2004) High performance discovery in time series: techniques and case studies. Springer, New York
5. Singh T, Kumari M (2021) Burst: real-time events burst detection in social text stream. J Supercomput. https://doi.org/10.1007/s11227-021-03717-4
6. Ebina R, Nakamura K, Oyanagi S (2011) A real-time burst detection method. In: 2011 IEEE 23rd international conference on tools with artificial intelligence, pp 1040–1046. https://doi.org/10.1109/ICTAI.2011.177
7. Kleinberg J (2002) Bursty and hierarchical structure in streams. In: 8th ACM SIGKDD international conference on knowledge discovery and data mining. https://www.cs.cornell.edu/home/kleinber/bhs.pdf. Accessed 09 June 2021

8. Tattershall E, Nenadic G, Stevens RD (2020) Detecting bursty terms in computer science research. Scientometrics 122:681–699. https://doi.org/10.1007/s11192-019-03307-5

9. Aggarwal CC, Subbian K (2012) Event detection in social streams. In: Proceeding 2012 SIAM international conference data mining, pp 624–635

10. Carbonell JG, Yang Y, Laferty J, Brown R, Pierce T, Liu X (1999) CMU Approach to TDT-2: segmentation, detection, and tracking. In: Proceedings of the 1999 DARPA broadcast news conference. https://doi.org/10.1184/R1/6604133.v1. Accessed 11 June 2021

11. Lee P, Lakshmanan LV, Milios EE (2014) Incremental cluster evolution tracking from highly dynamic network data. In: 30th International conference on IEEE data engineering (ICDE), pp 3–14

12. Orr W, Tadepalli P, Fern X (2018) Event detection with neural networks: a rigorous empirical evaluation. In: Proceedings of the 2018 conference on empirical methods in natural language processing. Association for Computational Linguistics, Brussels, pp 999–1004

13. McMinn AJ, Jose JM (2015) Real-time entity-based event detection for twitter. In: International conference of the cross-language evaluation forum for European languages, pp 65–77

14. Guille A, Favre C (2015) Event detection, tracking, and visualization in twitter: a mention-anomaly-based approach. Soc Netw Anal Min 5(1):18

15. He Q, Chang K, Lim E-P (2007) Using burstiness to improve clustering of topics in news streams. In: ICDM '07: Proceedings of the 2007 seventh IEEE international conference on data mining. IEEE Computer Society, Washington, DC, pp 493–498

16. He Q, Chang K, Lim E-P, Zhang J (2007) Bursty feature representation for clustering text streams. In: Proceedings of the seventh SIAM international conference on data mining, Minneapolis, Minnesota, pp 491–496

17. Lappas T, Arai B, Platakis M, Kotsakos D, Gunopulos D (2009) On burstiness-ware search for document sequences. In: Proceedings of the 15th AC, SIGKDD international conference on knowledge discovery and data mining, New York, pp 477–486

18. Sakkopoulus E, Antoniou D, Adamopoulou P, Tsirakis N, Tsakalidis A (2010) A web personalizing technique using adaptive data structures: the case of bursts in web visits. J Syst Softw 83:2200–2210

19. Kumar R, Novak J, Raghavan P, Tomkins A (2005) On the bursty evolution of blogspace. World Wide Web 8:159–178. https://doi.org/10.1007/s11280-004-4872-4

20. Platakis M, Kotsakos D, Gunopulos D (2008) Discovering hot topics in the blogosphere. In: Proceedings of the 2nd Panhellenic scientific student conference on informatics, related technologies and applications, Samos, pp 122–1332

21. Weng J, Lee B-S (2011) Event detection in twitter. In: Fifth international AAAI conference on weblogs and social media. https://www.aaai.org/ocs/index.php/ICWSM/ICWSM11/paper/view/2767/3299 Accessed 21 Feb 2021

22. Diao Q, Jiang J, Zhu F, Lim EP (2012) Finding bursty topics from microblogs. In: Proceedings of the 50th annual meeting of the association for computational linguistics: long papers-volume 1, ACL '12, pp 536–544

23. Mathioudakis M, Koudas N (2010) Twittermonitor: trend detection over the twitter stream. In: Proceedings of the 2010 ACM SIGMOD international conference on management of data, SIGMOD '10, pp 1155–1158

24. Xie S, Wang G, Lin S, Yu PS (2012) Review spam detection via temporal pattern discovery. In: Proceedings of the 18th ACM SIGKDD international conference on Knowledge discovery and data mining - KDD '12. ACM Press, Beijing, p 823

25. Fung GPC, Yu JX, Yu PS, Lu, H (2005) Parameter free bursty events detection in text streams. In: Proceedings of the 31st international conference on very large data bases, VLDB '05, pp 181–192

26. Takahashi Y, Utsuro T, Yoshioka M, Kando N, Fukuhara T, Nakagawa H, Kiyota Y (2012) Applying a burst model to detect bursty topics in a topic model. In: Isahara H, Kanzaki K (eds) Advances in natural language processing, Berlin, pp 239–249

27. Pollack J, Adler D (2015) Emergent trends and passing fads in project management research: a scientometric analysis of changes in the field. Int J Proj Manag 33:236–248. https://doi.org/10.1016/j.ijproman.2014.04.011
28. He D, Parker DS (2010) Topic dynamics: an alternative model of bursts in streams of topics. In: Proceedings of the 16th ACM SIGKDD international conference on knowledge discovery and data mining, pp 443–452
29. Mane KK, Börner K (2004) Mapping topics and topic bursts in PNAS. Proc Natl Acad Sci USA 101:5287–5290. https://doi.org/10.1073/pnas.0307626100
30. Binder J (2015) Bursts. https://cran.r-project.org/web/packages/bursts/bursts.pdf. Accessed 13 June 2021

Chapter 7
Sentiment Analysis

Abstract Sentiment or opinion analysis employs natural language processing to extract a significant pattern of knowledge from a large amount of textual data. It examines comments, opinions, emotions, beliefs, views, questions, preferences, attitudes, and requests communicated by the writer in a string of text. It extracts the writer's feelings in the form of subjectivity (objective and subjective), polarity (negative, positive, and neutral), and emotions (angry, happy, surprised, sad, jealous, and mixed). Thus, this chapter covers the theoretical framework and use cases of sentiment analysis in libraries. The chapter is followed by a case study showing the application of sentiment analysis in libraries using two different tools.

7.1 What Is Sentiment Analysis?

Sentiment analysis (also referred to as subjectivity analysis or opinion mining or emotion artificial intelligence) is a natural language processing (NLP) technique that identifies important patterns of information and features from a large text corpus. It analyzes thought, attitude, views, opinions, beliefs, comments, requests, questions, and preferences expressed by an author based on emotion rather than a reason in the form of text towards entities like services, issues, individuals, products, events, topics, organizations, and their attributes. It finds the author's overall emotion for a text where text can be blog posts, product reviews, online forums, speech, database sources, social media data, and documents. It usually consists of three elements depending on the context:

1. **Opinions or emotions**: An opinion is also referred to as polarity, whereas emotions can be qualitative such as sad, joy, angry, surprise, disgust, or happy or quantitative such as rating a movie on a scale of one to ten.
2. **Subject**: It refers to the subject of the discussion where one opinion can discuss more than one aspect of the same subject, for instance, the camera of the phone is great, but the battery life is disappointing.
3. **Opinion holder**: It refers to the author/person who expresses the opinion.

March 17 2020 the journal Nature Medicine provided strong evidence against the "engineered in a lab" idea. The study found a key part of SARS-CoV-2, known as the spike protein, would almost certainly have emerged in nature and NOT as a lab creation!

a) Positive tweet

Hi Mike what do you make of this paper? So much different info out there would like to get your take. Thanks.

b) Neutral tweet

One year since one of the most flawed papers on the origins of SARS-CoV-2 was published......

c) Negative tweet

Fig. 7.1 Example showing different sentiments for a particular research paper. (**a**) Positive tweet. (**b**) Neutral tweet. (**c**) Negative tweet

Texts are thus categorized as *subjective* if they reflect opinion; *objective* if they express a fact; *positive* if they present a state of satisfaction, bliss, or happiness; *negative* if they present a state of dejection, disappointment, or sorrow; or *neutral* if they present a state that is neither negative nor positive. It classifies the author's feelings into polarity (positive, negative, or neutral) and subjectivity (objective or subjective). Polarity is measured on a scale of -1 to $+1$, where -1 means very negative, 0 means neutral, and $+1$ means very positive. On the other hand, subjectivity is measured on a scale of 0–1, where 0 means very objective, and 1 means very subjective. Sentiment analysis involves tasks such as sentiment classification, subjectivity classification, summarization of opinions, and sentiment extraction, among others. Figure 7.1 shows an example of three tweets segregated based on their sentiments for the same research paper.

7.1.1   Levels of Granularity

Sentiment analysis can be performed at the following levels of granularity:

1. **Document Level**: It looks at the whole document and tags each document with its sentiment. It finds polarity and subjectivity of each sentence or word and then combines them to find the polarity and subjectivity of the document where it

assumes that the individual document emphasizes a single object and consists of opinion from a single author.

2. **Sentence or Phrase Level**: It determines and tags opinions expressed in each sentence with their polarity and subjectivity. It determines the sentiment orientation for each word in the phrase or sentence and then merges them to get the sentiment of the entire phrase or sentence.

3. **Aspect or Feature Level**: It labels each word with its sentiment and identifies the entity towards which the sentiment is directed using techniques like dependency parsing. It first recognizes and extracts object feature that the opinion holder mentions; then finds if the opinion on feature was negative, neutral, or positive; and finally finds similar features.

4. **Word Level**: It uses the prior polarity of words at sentence and document levels and uses a dictionary- or corpus-based approach.

7.1.2 Approaches for Sentiment Analysis

7.1.2.1 Rule or Lexicon

It has a predefined list of words in a dictionary with valence scores manually created with rules by humans. The algorithms match the words from the lexicon to the words in the text and either sum or take the average scores for the total appearance of the sentiment for the sentence or document. It is quite fast but might fail at specific tasks because the polarity of words may change with the problem, which will not be reflected in a predefined dictionary. It can further be classified into:

1. Dictionary-based: Highly used words are collected and annotated manually, such as SentiWordNet, which used the terms from WordNet and assigned sentiment scores to them. The limitation of the dictionary-based method is that it cannot be used for domain- and context-specific orientations.

2. Corpus-based: It provides a dictionary for specific domains and tags sentiments to related words based on semantics such as WordNet or statistical technique such as latent semantic analysis (LSA).

7.1.2.2 Automatic or Supervised Machine Learning

It is an automated system based on machine learning (see Chap. 8) and heavily relies on historical labeled data with known sentiment to predict the sentiment of the new data (classification problem). It is quite powerful but might take a while to train the data. The most common features that are used in automatic sentiment classification are parts of speech, negations, opinion words and phrases, and term presence and their frequency. Figure 7.2 demonstrates the supervised machine learning process for sentiments.

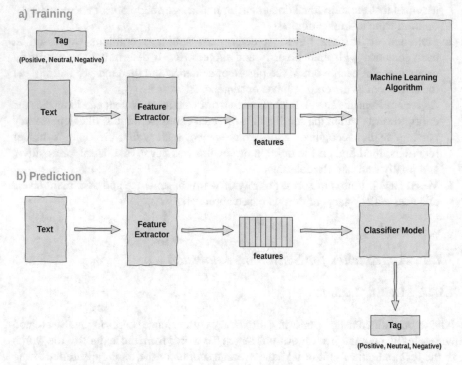

Fig. 7.2 Example showing the procedure of automatic tagging of sentiments. (**a**) Training. (**b**) Prediction

7.1.3   How to Perform Sentiment Analysis?

The typical steps to perform sentiment analysis are:

1. Pre-processing of data. It may include (i) removal of replies, mentions, URLs, hashtags, retweets, (ii) correction of spelling errors using a specific dictionary such as Hunspell dictionary, and (iii) replacing of abbreviations and shorthand notions using a specific dictionary such as SMS (Short Message Service) dictionary for social media data in addition to other pre-processing tasks covered in Sect. 3.3.
2. Feature extraction to extract the aspects from the processed data. This will be used to compute the polarity and subjectivity. This may include determining the n-gram with their frequency, part-of-speech tagging, identifying phrases and idioms, positioning of terms, negation, or syntactic patterns like collocations.
3. Performing sentiment analysis using an appropriate algorithm or open-source tool. The tools calculate the compound sentiment score for each article/post, giving an overall sentiment score.
4. Visualizing the sentiments.

There is a need to maintain order in a sentiment analysis process. Thus, a corpus is generally used instead of the bag-of-words (BOW) approach as it will fail to handle comparison and entity recognition. It is essential to determine sentiment analysis at a fine grain as the author may not like or dislike everything about a product or service and can have a point of view for several aspects that can be positive or negative.

7.1.4 Available Tools and Packages

There are various open-source tools such as SentiStrength,[1] LightSide.[2] Sentiment Viz,[3] NodeXL,[4] AYLIEN,[5] MonkeyLearn,[6] RapidMiner,[7] and Orange[8] available for non-programmers to perform sentiment analysis. Some of the popular packages in R to perform sentiment analysis include syuzhet,[9] SentimentAnalysis,[10] mscstexta4r,[11] sentimentr,[12] and quanteda,[13] whereas some popular packages to perform sentiment analysis in Python include NLTK,[14] NLP Architect,[15] VADER,[16] Polyglot,[17] TextBlob,[18] Flair,[19] and CoreNLP.[20] Most of the tools mentioned above are covered in detail in Chap. 10.

---

[1] http://sentistrength.wlv.ac.uk/.

[2] http://www.cs.cmu.edu/cprose/LightSIDE.html.

[3] https://www.csc2.ncsu.edu/faculty/healey/tweet_viz/tweet_app/.

[4] https://nodexlgraphgallery.org/Pages/AboutNodeXL.aspx.

[5] https://aylien.com/blog/using-entity-level-sentiment-analysis-to-understand-news-content.

[6] https://monkeylearn.com/sentiment-analysis-online/.

[7] https://rapidminer.com/.

[8] https://orangedatamining.com/.

[9] https://cran.r-project.org/web/packages/syuzhet/index.html.

[10] https://cran.r-project.org/web/packages/SentimentAnalysis/index.html.

[11] https://cran.r-project.org/web/packages/mscstexta4r/index.html.

[12] https://cran.r-project.org/web/packages/sentimentr/index.html.

[13] https://cran.r-project.org/web/packages/quanteda/index.html.

[14] https://pypi.org/project/nltk/.

[15] https://intellabs.github.io/nlp-architect/sentiment.html.

[16] https://github.com/cjhutto/vaderSentiment.

[17] https://polyglot.readthedocs.io/en/latest/.

[18] https://pypi.org/project/textblob/.

[19] https://pypi.org/project/flair/.

[20] https://stanfordnlp.github.io/CoreNLP/index.html.

7.1.5   Applications

Sentiment analysis is commonly applied in the domain of reviews of customer services and products. It is usually performed to monitor customer feedback and understand their needs. Social media is the primary focal point of various sentiment analysis applications, for instance, tracking the reputation of a particular brand or real-time analysis of posts/tweets containing a particular query of interest. It also helps to present significant value to candidates running for different positions and helps the managers monitor how voters relate to their speeches, feel about various issues, and relate to the candidates' actions. Thus, sentiment analysis has wide applications in business, customer feedback, brand monitoring, reputation management, customer support, product analysis, market research, competitive research, voice of employee, voice of customer, financial marketing, and social media monitoring.

7.1.6   Advantages

1. Classifying data at a large scale based on their polarity
2. Real-time analysis
3. Comparatively few categories/attributes like polarity compared to text categorization
4. Having nondependent categories/attributes
5. Having a relation between topic, domain, and user as opposed to text categorization
6. Simple and efficient

7.1.7   Limitations

1. Complexity in determining the real meaning of the expressions expressed by the opinion holder.
2. Sarcasm, irony, and implication are common and hard to decipher.
3. Different words might have different polarity or subjectivity in different contexts.
4. Same sentence or phrases might have different meaning in different domains.
5. Sentiments can be negated in different ways and it is difficult to identify such negations.
6. Dependent on order.

7. Content-dependent opinion words cannot be processed.
8. Can result in under- or over-analyzed sentiments if the used dictionary is too sparse or exhaustive.

7.2   Sentiment Analysis and Libraries

Social media contains potentially high valuable insight into the thoughts and depositions of the general public. Many libraries turn to social media to involve their patrons in an online space as it offers real-time venues and channels of communication, information sharing for knowledge exchange, and interactive dialogue [1–6]. Various case studies on *library use* of different social media tools, viz., Pinterest, Flickr, YouTube, Facebook, and Twitter, have been studied. Several studies [5, 7–9] showed how Facebook could promote library services from a librarian's perspective. It has been shown that reference inquiries were mostly asked through Facebook when compared through phone, email, traditional face-to-face questions, instant messaging, and outside reference desk shifts [10]. In libraries, Facebook has been recognized for producing static hyperlinks to static library resources and community building, whereas Twitter has been used for communication and well-timed updates about current events and new resources [11, 12]. Sentiment analysis can help to identify influential players in the network of libraries on social media [13]. Libraries can leverage such rich data sources in making informed purchasing decisions and policies and gauge the public perception of the library's services. In libraries, sentiment analysis has been applied on library tweets [14–17], library Facebook posts[18], and tweets that might be of interest to subject librarians[19].

In libraries, sentiment analysis can be conducted on the text data retrieved from (i) social media platforms, (ii) digital or digitized survey feedback, (iii) live librarians' chat archive, (iv) reviews of books, or (v) comments on library management systems (LMS) to provide insight on the satisfaction level of the library users for the products and services offered and organized by the libraries which include workshops, seminars, and conferences. The success of any library directly depends on its patrons, so if patrons like the library's products and services, then the library is a success; otherwise the library needs to improve the products and services by making changes to them. To know if a library's products or services are successful or not, you need to analyze its patrons and one of the ways to analyze them is through sentiment analysis. Sentiment analysis computationally identifies and categorizes opinions from texts. By automatically analyzing users' survey feedback responses or their social media conversation, libraries can learn about the services and products that make their patrons happy or frustrated so that they can tailor new or modify their services and products to meet their customer's needs.

Vinit Kumar is an Assistant Professor at the Department of Library and Information Science, Babasaheb Bhimrao Ambedkar University, Lucknow, who performed the following study in association with **Maya Deori** (Research Scholar) and **Manoj Kumar Verma** (Associate Professor) from the Department of Library and Information Science, Mizoram University, Aizawl.

Story: Analysis of YouTube Video Contents on Koha and DSpace and Sentiment Analysis of Viewers' Comments

In the last decade, we have seen the adoption of the latest technological innovations in the services provided by the library and information centers. The library professionals have adopted tools ranging from integrated library management software for managing housekeeping operations to developing institutional repositories using digital content management software. Although these tools have been part of the curriculum of most library schools in India, library professionals struggle to learn the intricacies of these tools when they try to implement them practically. Professionals attempt to learn these tools by attending workshops and try to self-learn using the documentation of these tools. With the growing availability and abundance of tutorial videos about these tools on social media websites such as YouTube[a], professionals now have not only the opportunity to learn the features of these tools but also a chance to share their views and help each other in troubleshooting any difficulties they face while implementing these tools.

We commenced a project to evaluate the characteristic features and opinion mining of the comments regarding the contents of videos uploaded on two tools top-rated among the library professionals of South Asian countries. We chose Koha as a representative for integrated library management software and DSpace for digital library software. Since there is no default guideline checking the quality and relevancy of the contents posted on YouTube, we analyzed the relevance and content quality satisfaction of the viewers by analyzing the sentiments expressed in the comments section of the videos.

We collected the metadata of the videos on the two selected tools using Webometric Analyst[b]. Using the same software, we further extracted the comments on each relevant video and exported the metadata and comments in comma-separated value files. We further analyzed the dataset for the characteristic features of the videos using Google Sheets. For the text analysis of the sentiments, emotions, and intentions of the comments, we used Text Analysis API[c] from ParallelDots[d] that provides a commercial API for text analysis based on machine learning algorithms. We created a free account and deployed the Google Sheets add-on[e] supplied by the ParallelDots developers. The add-on returned the sentiments and emotions for each row of the Google Sheet data, which was further analyzed and visualized. One of the main reasons for using an API rather than an established implementation

(continued)

of a text mining algorithm using some programming language was the seamless integration with Google Sheets and easy implementation. We further calculated the word frequency and other visualizations of the terms from the comments using Orange[f].

We faced some difficulties while conducting this project, such as the dependability of the software Webometric Analyst to extract metadata about videos relating to Koha and DSpace. We could never confirm whether any videos were omitted from the retrieved set of videos as in manual searching; the retrieval results on YouTube are customized based on the preferences of the user searching for the videos. Similarly, the limitation for evaluating the number of comments in ParallelDots API for a free account is 1000, causing us to create a part-by-part analysis of the dataset deploying multiple free accounts. For analysis of a larger dataset, one has to buy any of the paid plans. Another limitation of the project was that since we deployed ParallelDots API, being a proprietary service, we could not evaluate the algorithm and had to depend on the output received from the API.

We assume that this project will be of use to the library professionals who are eager to learn and make informed decision about the relevance and quality of the contents of the videos on both Koha and DSpace and also to the professionals who are video content creators so they can get an overview of the content they are uploading and get feedback for their improvement. Further, there is a scope to conduct similar studies using sophisticated software on text mining.

We published the results of the project as a research article that may be cited as Deori, M., Kumar, V., and Verma, M.K. (2021), "Analysis of YouTube video contents on Koha and DSpace, and sentiment analysis of viewers' comments," Library Hi Tech. https://doi.org/10.1108/LHT-12-2020-0323.
References

[a] https://www.youtube.com
[b] http://lexiurl.wlv.ac.uk/
[c] https://komprehend.io/
[d] https://www.paralleldots.com/
[e] https://komprehend.io/add-on
[f] https://orangedatamining.com/

7.2.1   Use Cases

7.2.1.1   Scite

Scite[21] is a service based on citations that provides citation context and classifies them as supporting or contrasting. It is a sentiment analysis-based service that uses a deep learning model to automatically classify each citation.

7.2.1.2   Virtual Librarian Chat

Sentiment analysis of virtual chat will help to determine users' satisfaction level with the services provided by the library and its staff [20].

7.2.1.3   Altmetrics

Sentiment analysis can help to analyze the *qualitative* altmetric data to characterize the sentiments or feelings of the people sharing their thoughts about a particular topic. Thus, one can analyze authors' sentiments for posts on common themes and identify outlets that publish more positive or negative news/articles/posts. Analyzing sentiment analysis of social media data related to research papers at scale can give a sense of what people think about research and how they are engaging with it [21–23].

7.2.1.4   Marketing

Lamba and Madhusudhan [19] emphasized in their paper that "social media mining (SMM) provides a dynamically active platform to connect with libraries' users (both active and potential) to market the library's products and services." They suggested that "in today's digital environment, librarians should analyze user-generated content over social media for decision-making; understanding their users' needs; making predictions; conducting a voluntary survey and analyzing users' opinions towards a particular issue, product or service." They further introduced "a new way of conducting a *SWOT analysis* for libraries by performing sentiment analysis on library's social media data where positive posts can be considered as strengths; negative posts can be treated as weaknesses; the problems expressed by their users can be viewed as the opportunities; the sentiment analysis of other libraries which they view as competitors can be viewed as the threats posed to the functionality and utilization of the library" [19].

---

[21] https://scite.ai/.

7.3   Case Study: Sentiment Analysis of Documents Using Two Different Tools

7A: RapidMiner

About the Case Study This case study has been adapted from Lamba and Madhusudhan [19] to illustrate a sentiment analysis using RapidMiner.

Problem If you have text data such as tweets, virtual chat with a librarian, Facebook posts, and book reviews and want to provide temporal information service based on the topics trending on social media.

Goal   To identify the sentiments and emotions of the public.

About the Tool   Refer to Chap. 10, Sect. 10.2.3, to know more about `RapidMiner`.

Theory   "Mining online opinion is a form of sentiment analysis that is treated as a difficult text classification task. It analyzes emotions, opinions, comments, views, beliefs, questions, requests, preferences, and attitudes expressed by the author in the form of text and determines the overall essence of the text expressed by the author" [19].

Methodology Figure 7.3 shows the workflow of sentiment analysis used in the RapidMiner platform. For retrieving data from Twitter, the API and Key were obtained by creating an app at the Twitter Developer[22] site and were then validated into the Twitter operator of RapidMiner (Fig. 7.4). Similarly, the API ID and Key were obtained from AYLIEN's Developer Portal[23] and were then validated into the AYLIEN operator of RapidMiner (Fig. 7.5).

Twenty bi-gram queries "related to the facets of productivity were searched on Twitter from 9th February 2018 to 21st February 2018 for a *recent or popular* type of tweets with a limit of 1500 tweets per query in the English language" [19]. The mining of tweets was followed by performing sentiment analysis using the AYLIEN operator. Therefore, 20 different Excel files[24] were downloaded from RapidMiner after each query search that consisted of the tweet, polarity (negative, positive, or neutral), and subjectivity (objective or subjective).

Results This study helped to comprehend the sentiment and emotions expressed by the public on Twitter for 20 different queries related to productivity. 6416 tweets and retweets were identified for 13 days' period. The following observations were made (Fig. 7.6):

---

[22] https://developer.twitter.com/en/apps.

[23] https://aylien.com/.

[24] https://github.com/textmining-infopros/chapter7/blob/master/7a_processed_dataset.rar.

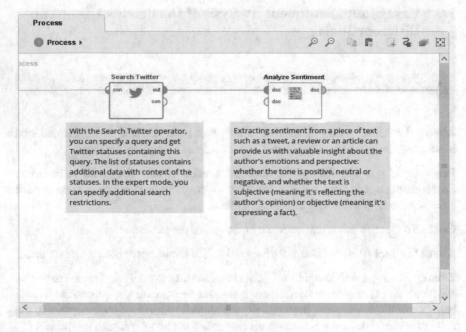

Fig. 7.3 Screenshot showing the workflow of sentiment analysis

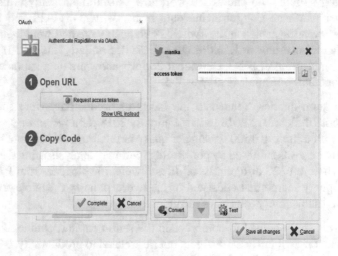

Fig. 7.4 Screenshot showing authentication in Twitter operator

- *Digital productivity* was the most popular facet, whereas *managerial productivity* was the least popular facet.
- Tweets for *low productivity* facet was the most negative, whereas *material productivity* facet had the most positive tweets.
- Tweets for Asian productivity facet were most neutral, whereas *material productivity* facet had the least neutral tweets.
- Tweets for *fish productivity* facet was most subjective, whereas *crop productivity* facet was the most objective.

7B: R

Problem If you have text data such as tweets, virtual chat with a librarian, Facebook posts, and book reviews and want to determine the satisfaction level of the users with the provided services and products.

Goal To identify the sentiments and emotions of the users.

Prerequisite Familiarity with the R language.

Virtual RStudio Server You can reproduce the analysis in the cloud without having to install any software or downloading the data. The computational environment runs using BinderHub. Use the link (https://mybinder.org/v2/gh/textmining-infopros/chapter7/master?urlpath=rstudio) to open an interactive virtual RStudio environment for hands-on practice. In the virtual environment, open the `sentiment_analysis.R` file to perform sentiment analysis.

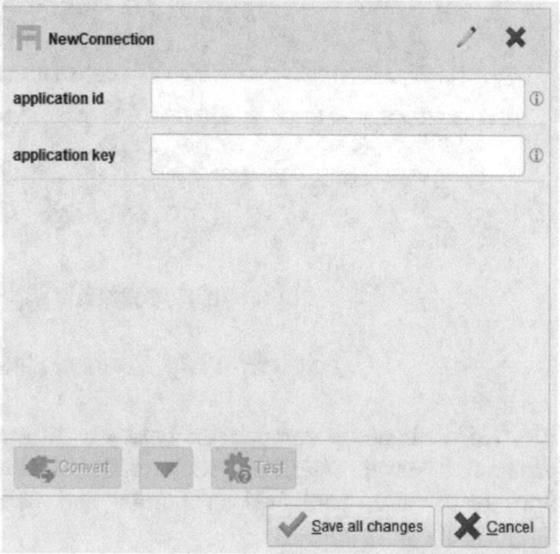

Fig. 7.5 Screenshot showing authentication in AYLIEN operator

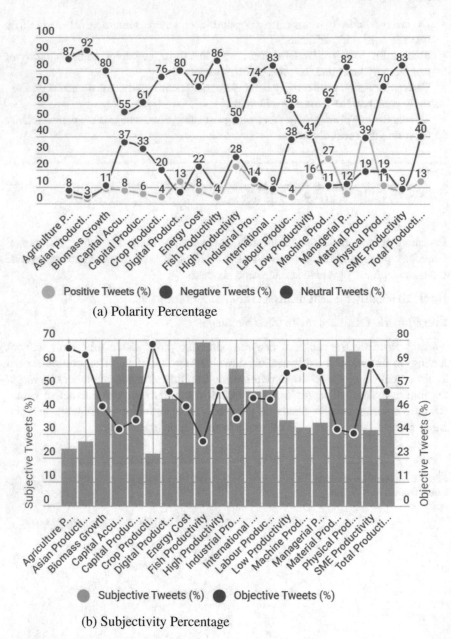

Fig. 7.6 Percentage comparison between different polarities and subjectivities for the studied productivity facets (©2018 Springer Nature, all rights reserved—reprinted with permission from Springer Nature, published in Lamba and Madhusudhan [19]). (**a**) Polarity percentage. (**b**) Subjectivity percentage

Virtual Jupyter Notebook You can reproduce the analysis in the cloud without having to install any software or downloading the data. The computational environment runs using BinderHub. Use the link (https://mybinder.org/v2/gh/textmining-infopros/chapter7/master?filepath=Case_Study_7B.ipynb) to open an interactive virtual Jupyter Notebook for hands-on practice.

Dataset A CSV file[25] containing 5000 book reviews web-scrapped from Amazon in 2018. The data is available in public domain at https://www.kaggle.com/shrutimehta/amazon-book-reviews-webscraped.

About the Tool Refer to Chap. 10, Sect. 10.2.1, to know more about R.

Theory The syuzhet package of R was used to perform sentiment and emotion analysis. It has four sentiment dictionaries to extract sentiment in addition to the sentiment extraction tool developed by the NLP group of Stanford.

Methodology and Results The libraries and the dataset required to perform sentiment analysis in R were loaded.

```
#Load libraries
library(syuzhet)
library(tm)
library(twitteR)

   #Load dataset
data<- read.csv("https://raw.githubusercontent.com/textmining
-infopros/chapter7/master/7b_dataset.csv")
```

The syuzhet package works only on vectors. So, the data was converted to a vector.

```
vector <- as.vector(t(data))
```

For each book review, scores were determined for different emotions, where a score with value 0 implies the emotion is not associated with the review, and the score of value 1 means there is an association between the emotion and the review. Subsequently, a higher score indicates stronger emotion.

---

[25] https://github.com/textmining-infopros/chapter7/blob/master/7b_dataset.csv.

Table 7.1 Sentiment score
for different sentiments

Sentiment score	Sentiment
< 0	Negative
= 0	Neutral
> 0	Positive

```
#Sentiment analysis
emotion.data <- get_nrc_sentiment(vector)
```

The following output gives a better representation of the book reviews with the associated emotions.

```
emotion.data2 <- cbind(data, emotion.data)
```

Sentiment scores were then computed for each book review using the built-in dictionary of the package that assigns sentiment score to different words. Table 7.1 shows the sentiment for the different range of sentiment scores.

```
sentiment.score <- get_sentiment(vector)
```

Reviews were then combined with both emotion and sentiment scores.

```
sentiment.data = cbind(sentiment.score, emotion.data2)
```

Positive, negative, and neutral reviews were then segregated and saved in three different CSV files.[26,27,28] Out of 5000 book reviews, 3587 were identified as positive, 1349 were identified as negative, and 64 were identified as neutral.

---

[26] https://github.com/textmining-infopros/chapter7/blob/master/positive_book_reviews.csv.

[27] https://github.com/textmining-infopros/chapter7/blob/master/negative_book_reviews.csv.

[28] https://github.com/textmining-infopros/chapter7/blob/master/neutral_book_reviews.csv.

```
#Getting positive, negative, and neutral reviews with associated scores
positive.reviews <- sentiment.data[which
(sentiment. data$sentiment.score > 0),]
write.csv(positive.reviews, "positive.reviews.csv")

negative.reviews <- sentiment.data[which
(sentiment.data$sentiment.score < 0),]
write.csv(negative.reviews, "negative.reviews.csv")

neutral.reviews <- sentiment.data[which
(sentiment.data$sentiment.score == 0),]
write.csv(neutral.reviews, "neutral.reviews.csv")
```

Lastly, a graph was plotted to visualize how the narrative is structured with the sentiments across the book reviews.

```
#Plot1: Percentage-Based Means
percent_vals <- get_percentage_values
(sentiment.score, bins=20)

plot(percent_vals,
type="l",
main="Amazon Book Reviews using Percentage-Based Means",
xlab="Narrative Time",
ylab="Emotional Valence",
col="red")
```

In Fig. 7.7, the x-axis presents the flow of time from start to end of the book reviews, and the y-axis presents the sentiments. In order to compare the trajectory of shapes, the text was divided into an equal number of chunks, and then the mean sentence

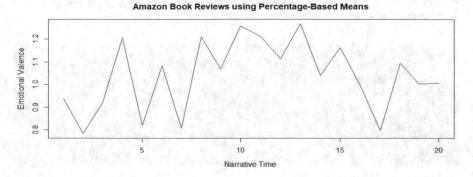

Fig. 7.7 Graph showing Amazon book reviews using percentage-based means

valence for each chunk was calculated. For this case study, the sentiments from the reviews were binned into 20 chunks where each chunk had 20 sentences.

The figure shows that the book reviews remain in the positive zone for all the 20 chunks. It dropped towards a comparatively less positive zone at many instances but never reached a neutral or negative zone. The limitation of the *percentage-based sentiment mean normalization* method is that in large texts, extreme emotional valence gets watered down, and the comparison between different books or texts becomes difficult.

To overcome the limitations of the *percentage-based sentiment mean normalization* method, the *discrete cosine transformation (DCT)* method was used as it gives an improved representation of edge values.

```
#Plot2: Discrete Cosine Transformation (DCT)
dct_values <- get_dct_transform(sentiment.score,
low_pass_size = 5,
x_reverse_len = 100,
scale_vals = F,
scale_range = T)

plot(dct_values,
type ="l",
main ="Amazon Book Reviews using Transformed Values",
xlab = "Narrative Time",
ylab = "Emotional Valence",
col = "red")
```

In Fig. 7.8, the x-axis presents the flow of time from start to end of the book reviews, and the y-axis presents the sentiments where 5 reviews were retained for low pass filtering, and 100 were returned. Figure 7.8 shows the transformed graph from the percentage-mean method. The reviews were of negative valence at the beginning that changed to positive valence and again dropped towards negative valence.

Moreover, eight different emotions, viz., anticipation, trust, joy, sadness, surprise, fear, anger, and disgust, were visualized using a bar plot.

```
#Plot3: Emotions Graph
barplot(sort(colSums(prop.table(emotion.data[, 1:8])))),
horiz=TRUE,
cex.names=0.7,
las=1,
main="Emotions in Amazon Book Reviews",
xlab = "Percentage")
```

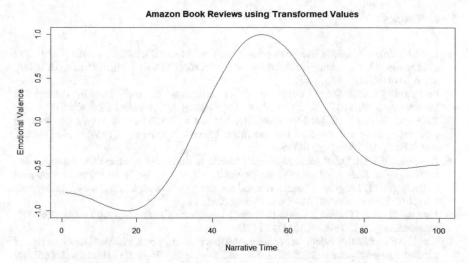

Fig. 7.8 Graph showing Amazon book reviews using discrete cosine transformation (DCT)

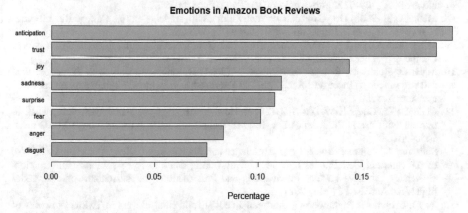

Fig. 7.9 Graph showing emotions for Amazon book reviews

Figure 7.9 shows that more than 15% of reviewers showed anticipation and trust emotion in their book reviews. Around 10% of the reviewers showed sadness, surprise, and fear emotions in their reviews, whereas the remaining reviewers showed anger and disgust.

References

1. Palmer S (2014) Characterizing university library use of social media a case study of Twitter and Facebook from Australia. J Acad Librarianship 40(6):611–619. https://doi.org/10.1016/j.acalib.2014.08.007
2. Burkhardt A (2010) Social media: a guide for college and university libraries. College Res Libraries News 71(1):10–24. https://crln.acrl.org/index.php/crlnews/article/view/8302
3. Collins G, Haines L (2012) Measuring libraries' use of YouTube as a promotional tool: an exploratory study and proposed best practices. J Web Librarianship 6(1):5–31. http://dx.doi.org/10.1080/19322909.2012.641789
4. Aziz NA, Chia YB, Loh H (2010) Sowing the seeds: towards reaping a harvest using social web applications in Nanyang Technological University Library. In: World library and information congress: 76th IFLA general conference and assembly, Gothenburg. https://www.ifla.org/past-wlic/2010/245-aziz-en.pdf. Accessed 24 May 2021
5. Aharony N (2012) Facebook use in libraries: an exploratory analysis. Aslib Proc 64:358–372. https://doi.org/10.1108/00012531211244725
6. Kim Y, Abbas J (2010) Adoption of Library 2.0 functionalities by academic libraries and users: a knowledge management perspective. J Acad Librarianship 36(3):211–218. http://dx.doi.org/10.1016/j.acalib.2010.03.003
7. Landis C (2007) Social networking sites: getting friendly with our users. College Res Libraries News 68(11):709–712. https://doi.org/10.5860/crln.68.11.7907
8. Matthews B (2006) Do you Facebook: networking with students online. College Res Libraries News 67(5):306–307. https://doi.org/10.5860/crln.67.5.7622
9. Miller SE, Jensen LA (2007) Connecting and communicating with students on Facebook. Comput Libraries 27(8):18–22
10. Mack D, Behler A, Roberts B, Rimland E (2007) Reaching students with Facebook: data and best practices. Electron J Acad Special Librarianship 8(2). https://digitalcommons.unl.edu/ejasljournal/85/
11. Chen DY-T, Chu SK-W, Xu S-Q (2012) How do libraries use social networking sites to interact with users. Proc Amer Soc Inf Sci Technol 49:1–10. https://doi.org/10.1002/meet.14504901085
12. Salisbury L, Laincz J, Smith J (2012) Science and Technology Undergraduate Students' Use of the Internet, Cell Phones and Social Networking Sites to Access Library Information. University Libraries Faculty Publications and Presentations. https://scholarworks.uark.edu/libpub/10. Accessed 24 May 2021
13. Yep J, Brown M, Fagliarone G, Shulman J (2017) Influential players in Twitter networks of libraries at primarily undergraduate institutions. J Acad Librarianship 43:193–200. https://doi.org/10.1016/j.acalib.2017.03.005
14. Lund BD (2020) Assessing library topics using sentiment analysis in R: a discussion and code sample. Public Serv Quarterly 16(2):112–123. https://doi.org/10.1080/15228959.2020.1731402
15. Patra SK (2019) How Indian libraries tweet? Word frequency and sentiment analysis of library tweets. Ann Library Inf Stud 66(4):131–139. http://op.niscair.res.in/index.php/ALIS/article/view/26636/465477307
16. Al-Daihani SM, Abrahams A (2016) A text mining analysis of academic libraries' Tweets. J Acad Librarianship 42(2):135–143. https://doi.org/10.1016/j.acalib.2015.12.014
17. Stewart B, Walker J (2018) Build it and they will come? Patron engagement via Twitter at historically black college and university libraries. J Acad Librarianship 44:118–124. https://doi.org/10.1016/j.acalib.2017.09.016
18. Al-Daihani SM, Abrahams A (2018) Analysis of academic libraries' Facebook posts: text and data analytics. J Acad Librarianship 44:216–225. https://doi.org/10.1016/j.acalib.2018.02.004

19. Lamba M, Madhusudhan M (2018) Application of sentiment analysis in libraries to provide temporal information service: a case study on various facets of productivity. Soc Netw Anal Min 8:63. https://doi.org/10.1007/s13278-018-0541-y
20. Logan J, Barrett K, Pagotto S (2019) Dissatisfaction in chat reference users: a transcript analysis study. College Res Libraries 80:925. https://doi.org/10.5860/crl.80.7.925
21. Friedrich N, Bowman TD, Stock WG, Haustein S (2015) Adapting sentiment analysis for tweets linking to scientific papers. In: 15th international society of scientometrics and informetrics conference (ISSI 2015), Istanbul. https://arxiv.org/abs/1507.01967. Accessed 21 May 2021
22. Hassan S-U, Saleem A, Soroya SH, et al (2020) Sentiment analysis of tweets through altmetrics: a machine learning approach. J Inf Sci 0165551520930917. https://doi.org/10.1177/0165551520930917
23. Hassan S-U, Aljohani NR, Tarar UI, et al (2020) Exploiting Tweet Sentiments in Altmetrics Large-Scale Data. https://arxiv.org/abs/2008.13023. Accessed 21 May 2021

Chapter 8
Predictive Modeling

Abstract This chapter covers a comprehensive theoretical framework for predictive modeling (or supervised machine learning). It also covers various biases, challenges, solutions, and use cases of predictive modeling in libraries. A case study that shows how library professionals can use predictive modeling to index/tag future textual resources without repeating a text mining technique, again and again, is also included.

8.1 What Is Predictive Modeling?

Traditionally, humans analyzed the data, but the volume of data surpasses their ability to make sense of it, which made them automate systems that can learn from the data and the changes in data to adapt to the shifting data landscape. Machine learning uses statistical algorithms to learn from examples and is about creating statistical programs called *models*. A model takes input based on lots of example data (such as photos, webpages, journal articles, speeches, social media profiles) to train the model and give an output. In statistics, independent variables are referred to as input variables, whereas dependent variables are referred to as outcome variables. In contrast, input or independent variables are referred to as *features*, and output or dependent variables are referred to as target values or labels in a machine learning (ML) process. In supervised machine learning or predictive modeling, the objective is to create predictive methods that can accurately predict the outcomes (target values or target labels) for new data.

Figure 8.1 shows the basic workflow of ML, which is an iterative (cyclic) process. Firstly, we convert the research problem into a representation that the computer can process, which involves conversion of each object, often called sample, into a set of *features* that describe the object. Then, we make an initial guess of what good *features* are present for the research problem and the classifier that might be appropriate for the model. Secondly, we pick a learning model (typically the type of classifier you want to learn), train the model using the training data, and produce some evaluation to see how well the classifiers worked. Finally, we perform a failure

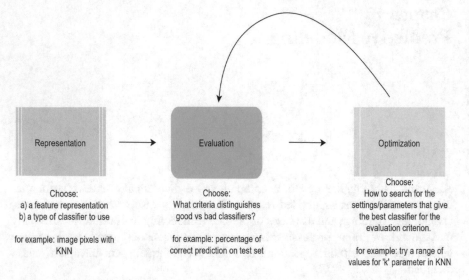

Fig. 8.1 Diagram of basic machine learning workflow

analysis based on the examples that were not classified correctly to see where the system is still making a mistake.

In the early 1990s, ML was applied in text categorization (also referred to as document classification) studies. Document classification is the process of assigning documents to one or more defined categories or classes. It can be applied using supervised learning-based classification that requires training of data or unsupervised learning-based clustering that does not require training of data. It consists of three components: (i) *document collection* divided into training and test dataset; (ii) *classifier*, which is a mathematical function and implements classification; and (iii) *features* such as bag-of-words or TF-IDF. Various applications of ML include:

- Translation software
- Speech recognition
- Image recognition
- Online shopping
- Recommending sites based on what the users like or watched in the past, and patterns or preferences related to the users
- Fraud detection
- Web searching through query spell-checking, result ranking, content classification and selection, and advertisement placement
- Handwriting recognition and categorization

8.1.1   Why Use Machine Learning?

In most computing problems that need to be solved, a script is written manually that further includes a series of steps to solve that particular problem. However, for some problems like speech recognition, it is difficult to write a program manually. ML is a technique that helps to learn complex rules efficiently. It uses labeled instances called *training* data instead of programming all the rules manually. Thus, it provides a more accurate and flexible way to solve a complex problem.

ML examines how computers can program a complex problem by themselves and improve their performance automatically as they acquire more practice of learning the data using the training phase of the model. This experience can take various forms of data or situations such as labeled instances like tagged documents or clicking on the search page in a search engine by users. ML relies on algorithms to infer predictions by learning from the input data and also measures how sound those predictions are. Thus, the goals and features of ML include:

- Training of the best model in the least amount of time
- Reducing human effort and the expertise required in ML
- Improving the performance of ML models
- Increasing reproducibility and establishing a baseline for scientific research or applications

8.1.2   Machine Learning Methods

There are generally three types of ML methods: **supervised**, **unsupervised**, and **semi-supervised**. *Supervised* ML (also referred to as predictive modeling or predictive analytics) uses data (referred to as training) to answer questions (referred to as making predictions or inference) using a statistical model based on an ML algorithm. It is used to predict some output variable that is associated with each input item to make predictions. The model learns to predict target values from the labeled data. Thus, in supervised ML, we can train a model to make better and useful predictions for previously seen text data. In *unsupervised* ML, some kind of structure is being found in the unlabeled data. It identifies groups of similar examples (clustering) and unusual patterns (outlier detection) in the data.

In many cases of unsupervised learning, we have only input data and do not have any labels to go with the data. Unsupervised ML can be used for tasks like summarizing the input data, visualizing the structure, and grouping it into different categories. In *semi-supervised* ML, a large amount of unlabeled data is assigned for training along with a small amount of labeled data. It is generally preferred to increase your training set when you do not have enough labeled data and do not have the resources to get more data. It reduces the time to hand-label dataset and increases the productivity and efficiency to generate the training data. However, it generates less trustworthy outcomes than traditional supervised techniques, and there is no way to verify if the algorithm produced accurate labels.

8.1.3 Feature Selection and Representation

Predictive modeling for text data is a process of building a statistical model to estimate the output using a set of input of *features* (e.g., tokens, parts of speech). These *features* can be based on either the morphology (i.e., how words are formed) or syntax (how sentences are formed from words) of the language and help to transform the unstructured text into a mathematical representation that can be used in a predictive model. *Features* are generally created using parts-of-speech tagging, tokenization, sentiment analysis, and named entity recognition, where one or more *features* can be used in a predictive model for prediction.

There are so many words that are irrelevant in high-dimensional textual data, and the removal of such words is called *feature selection*. In contrast, *feature engineering* is the process of taking an object and converting it to a number that the computer can understand and is used in various ML algorithms to try to predict the label of an object. Removal of stopwords and inverse document weighting frequency (IDF) of words are forms of feature selection, whereas the creation of word embeddings and matrix factorization methods are examples of feature engineering.

8.1.4 Machine Learning Algorithms

Most ML algorithms are based on statistics and can be quite mathematical and complex. Some of the commonly used algorithms for text categorization and classification are decision trees, support vector machine (SVM), neural networks, and naïve Bayes. *Optimization* is the process of choosing the number that can be used in the model that gives the best results on the training data. Most ML algorithms use features from input data into mathematical functions called *models*. The details of the models are automatically chosen to give the best results on the training data (*optimization*). An effective ML algorithm helps to generalize and accurately predict new data that were not seen during the training phase of the model. Some of the significant ethical and technical considerations when using ML are:

- Could your ML model harm people, especially marginalized groups, and could accelerate the biases and harms that already exist?
- Does your ML model shift power, especially to majority groups?
- Does your ML model fail on outliers?
- Does your ML model have strong confidence about wrong answers?

8.1.4.1 Generalization, Overfitting, and Underfitting

Generalization is an algorithm's capability to correctly predict new data that was previously unseen by the model. When a supervised ML algorithm performs well on a held-out test set, then it shows its ability to generalize, but how can we know that the trained model will generalize well on new data? Generally, ML assumes that the unseen test set has properties similar to the training set. Thus, the models that performed well on the training set are expected to perform well on the test set, but that might not happen if the model is tuned specifically for the trained data resulting in *overfitting* of the data.

Complex models tend to *overfit* the available training data and are unlikely to generalize well on new instances. *Overfitting* models capture complex patterns by identifying local variations in the training set but fail to see the global patterns in the training set that will help them generalize on the new unseen test data. This results in a hopeless optimistic accuracy on the training data for the test data accuracy if the model is *overfitting*. It is essential to understand, detect, and avoid *overfitting* when applying a supervised ML algorithm, as it might have some severe consequences for real-world problems. On the other hand, the models that cannot do well on the training set, are too simple, and unlikely to generalize well are called *underfit*.

Figure 8.2 presents a 2D binary classification problem where each data example is shown by a circle (red or green) and where each data example is associated with

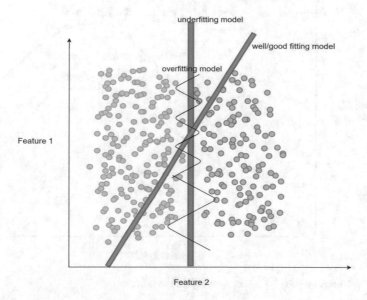

Fig. 8.2 Diagram showing an example of a classification problem

the two features. A *good or well fit* model (represented by blue line) finds the general difference between the positive and negative classes by identifying the global pattern and global separation between the two classes. An *overfitting* model (represented by black zigzag line) uses lots of parameters to capture the complex behavior and try to completely separate the green points from the red ones by forming a highly variable decision boundary. It captures and predicts the training data almost perfectly by locating too many local fluctuations but ignores the global trend as it does not have enough data for performance generalization on the new test data. An *underfitting* model (represented by the purple line) does not really look at the data to generalize and only looks at one feature. Therefore, it does not capture pattern and divides the classes in the training data in a simplistic model.

Relationship Between Model Complexity and Performance
The relationship between model complexity and model performance depends on whether the classifier performance is evaluated on the test set or the training set (Fig. 8.3). As a more complex model fits the training data, the model's accuracy moves upward. However, when the same trained model is measured on held-out test data, there is an initial increase in accuracy up to an optimal point due to the model's complexity, followed by a decrease in the accuracy of the test set due to the decrease in the model's complexity. This results in overfitting of the training

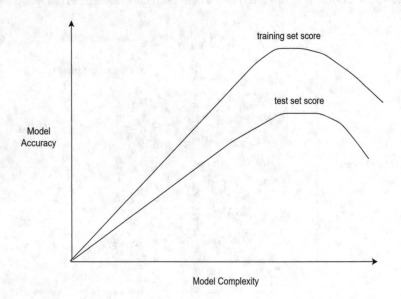

Fig. 8.3 Relationship between model accuracy and complexity

data and not capturing global patterns that help to generalize it to new data. Though this trend will vary from situation to situation, overall, this pattern remains the same across the supervising learning methods.

8.1.5   Classification Task

Some of the crucial tasks of supervised ML include classification and regression. As this chapter focuses on predictive modeling for text documents, it will only cover the classification task of ML. The ML task of classification is associated with a predictive modeling problem, which is used to predict a class label or group membership for a document. It can solve problems like recommendation systems or online search by predicting whether or not to display an advertisement or show a query suggestion or item on a page or suggest a product relevant to the user's query or what they have clicked or read in the past. There are mainly four types of classification task for supervised ML (Fig. 8.4):

1. **Binary Classification**: It classifies tasks with two class labels such as credit card fraud (fraudulent or not) to predict a Bernoulli probability distribution. It generally consists of one normal and one abnormal class, where 1 is assigned for the normal class and 0 for the abnormal class. Some famous binary classification algorithms are k-nearest neighbors, logistic regression, decision trees, SVM, and naïve Bayes. Classifiers like logistic regression and SVM do not support more than two classes and are designed for binary classification.
2. **Multi-Class Classification**: It classifies tasks with more than two class labels such as plant species classification to predict a Multinoulli probability distribution. It classifies the data to one of the classes amidst a range of known classes. Some of the popular algorithms for multi-class classification are decision trees, k-nearest neighbors, random forest, naïve Bayes, and gradient boosting.
3. **Multi-Label Classification**: It classifies tasks with two or more class labels and predicts one or more class labels for each label, such as photo classification (bicycle, person, fruit). It uses Bernoulli probability distribution to predict the outputs and uses a specialized version of a standard classification algorithm for multi-label classification, viz., multi-label gradient boosting, multi-label decision trees, and multi-label random forests.
4. **Imbalanced Classification**: It classifies tasks where instances for each class label are distributed unequally, such as outlier detection. It is similar to binary classification, where there is a majority of normal class instances and a minority of abnormal class instances in the training dataset. It requires special techniques such as random undersampling to undersample the majority class or SMOTE oversampling to oversample the minority class to change the composition of samples. It uses special algorithms such as cost-sensitive logistic regression, cost-sensitive support vector machines, or cost-sensitive decision trees to perform imbalanced classification.

Fig. 8.4 Scatterplots
showing different types of
classification. (**a**) Binary. (**b**)
Multi-class. (**c**) Multi-label.
(**d**) Imbalanced

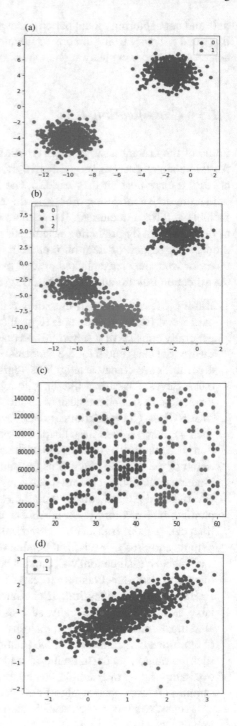

8.1.6 How to Perform Predictive Modeling on Text Documents?

In predictive modeling, a set of text documents are annotated or tagged with one set of the categories of the *features*. These documents are used to *train* a statistical model based on the ML algorithm. The untagged documents are then added to the trained model, which learns the patterns of different categories. Predictive models reflect the characteristics of their training data. Therefore, differences in language over time, between dialects, and various cultural contexts can prevent a model trained on a particular dataset from being applied to others. Thus, the primary motive of predictive modeling is to predict the *feature* of past/trained data on new/test data. Theoretically, there is no way to find out which algorithm will fit best for a particular problem. Instead, it is recommended to use controlled experiments and find which algorithm gives the best performance for a classification task based on their evaluation results. Hence, the steps to perform predictive modeling on text data are:

1. **Pre-processing**: The text documents are first pre-processed using tokenization, stopword removal, lemmatization, and stemming tasks (covered in Sect. 3.3).
2. **Feature Selection**: Features are then selected and engineered for the pre-processed data.
3. **Model Building**: A statistical model is built to run one or more algorithms on the dataset. The dataset is then divided into training and testing datasets. The model is first trained on the training dataset, followed by testing of data on the trained data to check how well the model predicts.
4. **Model Evaluation**: The model is evaluated using measures like precision, accuracy, f1 score, recall, or the area under the ROC curve (AUC). The model that fitted the data the best is then chosen.

8.1.6.1 Training and Testing Datasets

In ML, the data is usually split into a *training set* and a *test set*. The *training set* should always be bigger so that enough data is provided to the ML algorithm to learn effectively. A good split might be 4/5th for the *training set* and 1/5th for the *testing set*. ML can learn all the details of the training data and do well on it, but it probably will not do well on new data that was not in the training data. That is why ML practitioners keep some data just for testing. The test data is not used during the training phase. The percentage of correct classification on the test set will give a much better idea of how well the model will work in practice.

8.1.6.2 Model Evaluation

Model evaluation helps to provide critical feedback on the performance of the trained model. The evaluation step results may help to identify the data instances that are incorrectly predicted or classified and suggest different features or other refinements during the feature selection and model building phases. Evaluation measures help to choose between different trained models or configurations. As different ML applications have different goals, it is vital to select the evaluation metric that complements the application's objective. Therefore, one should select evaluation metrics for different models and then choose the model with the *best* results.

Supervised learning (continuous target) model for regression task can be evaluated using metrics such as mean square error (MSE), R2 score, or mean absolute error (MAE). In contrast, a supervised learning (discrete target) model, also referred to as a classification model, can be evaluated using metrics such as precision, f1 score, recall, accuracy, area under ROC curve (AUC), etc. A confusion matrix is usually used to represent the performance of a classifier (or classification model) on the test data with known true values. It helps to visualize the performance of the predictive model and identifies the mislabeled classes. A confusion matrix evaluates one particular classifier with a fixed threshold. In Fig. 8.5, the phrases, positive and negative, are used to expect the classifier's prediction. In contrast, the phrases, true and false, are used to observe the classifier's prediction. The correctly labeled classes are on the diagonal of the confusion matrix where the true class matches the predicted instances. When you add all the instances along the diagonal, you will get the total number of correctly predicted instances for all the classes.

	Class 1 Predicted	Class 2 Predicted
Class 1 Actual	TP	FN
Class 2 Actual	FP	TN

Positive (P): Observation is positive
Negative (N): Observation is not positive
True Positive (TP): Observation is positive, and is predicted to be positive
False Negative (FN): Observation is positive, but is predicted negative
True Negative (TN): Observation is negative, and is predicted to be negative
False Positive (FP): Observation is negative, but is predicted positive

Fig. 8.5 Example of confusion matrix

The performance of predictive models is usually judged using the following measures:

1. **Accuracy**: Accuracy is defined as the distribution of correct classification (true positive, true negative) from the total number of examples. It can be misleading. Therefore, precision and recall are the primarily utilized measures.

$$\text{Accuracy} = \frac{(TP + TN)}{(TP + TN + FP + FN)} \tag{8.1}$$

2. **Precision**: Precision is defined as the distribution of the total number of correct positive classifications (true positive) from examples that are predicted as positives.

$$\text{Precision} = \frac{(TP)}{(TP + FP)} \tag{8.2}$$

3. **Recall or Sensitivity or True Positive Rate (TPR) or Problem of Detection**: The recall is the distribution of the total number of correct positive classifications (true positives) from cases that are actually positives.

$$\text{Recall} = \frac{(TP)}{(TP + FN)} \tag{8.3}$$

4. **Specificity or False Positive Rate (FPR)**: It gives the ratio of all the negative examples that the classifier incorrectly identifies as positive.

$$\text{FPR} = \frac{(FP)}{(TN + FP)} \tag{8.4}$$

5. **Kappa**: Cohen's kappa statistics compare the observed accuracy with an expected accuracy (random chance). It is beneficial in multi-class and imbalanced classification problems as in those cases, accuracy, precision, or recall values do not provide the complete picture of the classifier's performance. It is always less than or equal to 1.

Kappa value	Agreement
< 0	No agreement
0 to 0.20	Slight agreement
0.21 to 0.40	Fair agreement
0.41 to 0.60	Moderate agreement
0.61 to 0.80	Substantial agreement
0.81 to 1	Perfect agreement

$$\text{Kappa} = \frac{(Observed\,Accuracy - Expected\,Accuracy)}{(1 - Expected\,Accuracy)} \tag{8.5}$$

6. **F-Measure**: It measures test data accuracy and determines the harmonic mean of recall and precision. The score reaches its best value at 1 showing perfect precision and recall, whereas 0 shows the worst.

$$\text{F-measure} = \frac{2 * (Recall * Precision)}{(Recall + Precision)} \tag{8.6}$$

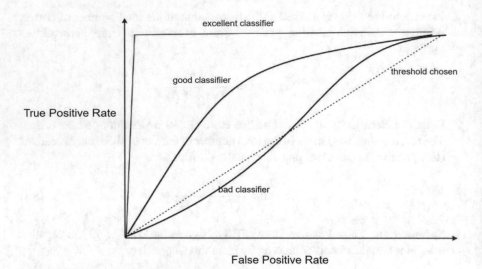

Fig. 8.6 ROC curve for different classifiers

7. **Receiver Operating Characteristic (ROC) Curve**: It examines the performance of a classifier (Fig. 8.6). It tests every possible threshold and plots each result as a point on the curve. It is a useful evaluative measure to test the classifier over a range of specificities/sensitivities. The area under curve (AUC) of ROC evaluates that classifier over all possible thresholds. An AUC value of 0 represents a very bad classifier, and an AUC value of 1 represents an optimal classifier.

So, for instance, for a search engine task, we might want a classifier with higher precision, but then the recall value of the classifier will decrease. This example illustrates a classic tradeoff that usually appears in ML operations. The increase in the precision of a classifier will result in a decrease in recall value, or the increase in the recall of a classifier will result in a decrease in precision. Recall-oriented ML tasks include searching and information extraction and are often aided by a human expert to filter out a false positive. In contrast, precision-oriented ML tasks include search engine ranking, query suggestion, document classification to annotate with topic tags, and other user-facing tasks.

Why Accuracy May Not Be a Good Evaluation Metric to Judge the Performance of a Classifier?
Classification problems with imbalanced classes are prevalent in ML. Suppose you have two classes: relevant (R) and nonrelevant (N), then out of 1000 randomly selected items, on average, one item is relevant and has an R label, and the rest of the 999 items are not relevant and labeled N. If you build a classifier to predict

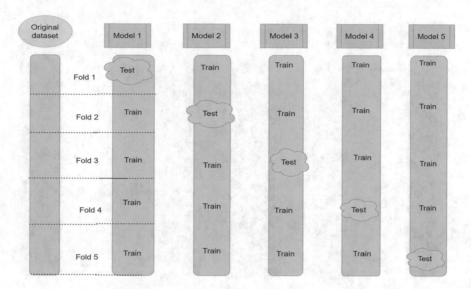

Fig. 8.7 Diagram showing five-fold cross-validation

the relevant items and see an accuracy of 99.9%, that is because the classifier did not look at all the instances and just blindly predicted the most frequent class (N, in our case). Thus, if you have a classifier that shows a high percentage of accuracy, then it could be a sign of *ineffective, erroneous, or missing features, poor choice of kernel or hyperparameter*, or *large class imbalance*. For an imbalanced classification problem, one should generally use metrics like AUC (area under the curve) and avoid using accuracy as the evaluation metric.

Cross-Validation
Cross-validation uses multiple train/test splits to train and evaluate separate models instead of evaluating a single train/test split for a single model. It provides a more stable and reliable estimate by averaging the performance of each training/test split. It is not used to tune a new model and only for the evaluation process. Figure 8.7 illustrates the function of cross-validation on a dataset. The most common type of cross-validation is the K-fold cross-validation, where K is commonly set to 5 or 10. The figure shows the partitioning of five equal (or close to equal) parts of data called *fold*, and each fold is trained in the five models. Model 1 is evaluated for fold 1 (test set) and trained using folds 2 through 5, whereas model 2 is evaluated for fold 2 (test set) and trained using folds 1, 3, 4, 5, and so on. Once the process is completed, five accuracy values will be generated and are reported as mean cross-validation scores. The advantage of performing multiple training/test splits is that it helps to determine the model's sensitivity for all cross-validation folds to check how bad or well the model will perform on the unseen test data. Thus, it helps to sort the best and worst performance estimates for the model. One problem can be that the data might have been created in analysis by K-fold cross-validation because the examples tend to

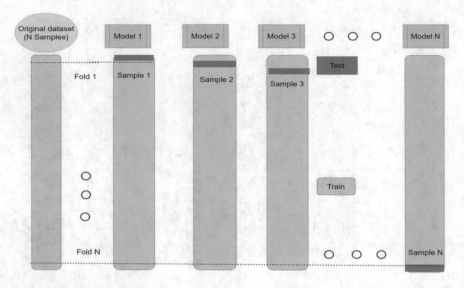

Fig. 8.8 Diagram showing five-fold cross-validation

show some biasness in the class label due to their ordering. This can be overcome by stratified K-fold cross-validation, where stratified folds contain a proportion of classes that match the overall dataset. All the classes are fairly represented in the test set. Another variation of K-fold cross-validation is the *leave-one-out cross-validation* (Fig. 8.8), where each fold comprises a sample (test set), and the rest of the data is used as the training set. It uses more computation but is reliable and provides an unbiased estimate of the model's performance. It gives the best result with a small dataset and can improve the model's estimate by maximizing the amount of the training data.

Model Selection Using Evaluation Metrics
By only using the test data or cross-validation to do model selection, we may get optimistic/overfitting generalization performance for the new unseen data. Instead, we should split the data into three sets: training set for *model building*, validation set for *model selection*, and test set for *model evaluation*. In practice, the training and test sets are created, followed by cross-validation on the training data for model selection where the test set is not seen until the end of the evaluation process, and the held-out test set is saved for final model evaluation.

8.1.7 Available Tools and Packages

Some of the popular open-source tools that can be used by non-programmers for ML are Teachable Machine,[1] WEKA,[2] RapidMiner,[3] and Orange.[4] The popular packages available in R for ML include h2o[5,6] tidyverse,[7] quanteda,[8] caret,[9] kernlab,[10] and Keras,[11] whereas the popular packages in Python are scikit-learn,[12] SciPy,[13] NLP Architect,[14] PyTorch,[15] and Keras.[16] Most of the tools mentioned above are covered in detail in Chap. 10.

8.1.8 Advantages

1. It collects or produces output data based on previous data/experience.
2. It discovers specific trends and patterns in the data that would not be evident to humans.
3. It is an automated process where machines can learn and make predictions to improve algorithms on their own.
4. The algorithms gain experience and keep getting improved accuracy and efficiency as more training data is added. This helps to make better decisions and make more accurate predictions.
5. It is good at handling multidimensional and multi-variety data.
6. It helps to optimize parameters and performance of the classifiers based on past data/experience.
7. It helps to solve real-world computational problems.

---

[1] https://teachablemachine.withgoogle.com/.

[2] https://www.cs.waikato.ac.nz/ml/weka.

[3] https://rapidminer.com/.

[4] https://orangedatamining.com/.

[5] https://docs.h2o.ai/h2o/latest-stable/h2o-r/docs/index.html.

[6] https://cran.r-project.org/web/packages/h2o/index.html.

[7] https://www.tidyverse.org/.

[8] https://quanteda.io/.

[9] https://cran.r-project.org/web/packages/caret/index.html.

[10] https://cran.r-project.org/web/packages/kernlab/index.html.

[11] https://keras.rstudio.com/.

[12] https://scikit-learn.org.

[13] https://www.scipy.org/.

[14] https://intellabs.github.io/nlp-architect/sentiment.html.

[15] https://pytorch.org/.

[16] https://keras.io/.

8.1.9   Limitations

1. It requires a large amount of unbiased and good quality training data.
2. It does not have the ability to accurately interpret results generated by algorithms.
3. Organization and rules of languages can be ambiguous that can affect the process of feature creation.
4. Language from a specific dialect often cannot be handled well with a model trained on data from the same language but not inclusive of that dialect.
5. The data trained on a specific set of texts (such as tweets) cannot be tested on other types (such as legal documents).
6. It is susceptible to errors, for instance, the model might get overtrained if the training set does not have examples that are needed in the class.
7. Good examples for each class are needed to train the classifier.
8. Training in supervised learning needs a lot of computational time.

8.2   Machine Learning and Libraries

ML methods and tools help to make library resources more usable for students, scholars, faculty, and the general public. It aids discoverability amidst information abundance. For instance, "imagine patrons browsing automatically-derived topics of interest across a digital library comprising thousands or millions of texts—more texts, certainly, than typical constraints on labor or expertise would allow us to imagine labelling manually" [1]. Many faculty and students seek to access rich data to train algorithms for research purposes. It becomes essential for librarians to assist them in finding the required datasets (training data).

Moreover, to make the library collection more accessible to the users, librarians need to push the vendors to get license access within a controlled environment to perform text and data mining, algorithmically based research, and machine learning on the content provided to them. In addition to that, as librarians build their own digital collections, they need to think both ethically and legally about how accessible they want to make their collection to users to perform machine learning and other text mining techniques. It is also essential to consider addressing questions related to the preservation of large datasets and the results from the ML projects such as (i) "Should we sustain both technical and intellectual access over the long term, and if so, how?", (ii) "How do we address the ethical and legal rights not only of the users of these datasets and the creators of scholarship but also of those represented in the datasets?", and (iii) "How do we support the replication of studies and the auditing of datasets, especially those used as training data in machine learning systems, for bias?" [2]. Many vendors are already providing services (for instance, TDM Studio[17] by Pro-Quest, Digital Scholar Workbench---Constellate[18] by JSTOR and Portico) to incorporate artificial intelligence, text mining, and machine learning into their platforms, services, and products (see Appendix 11.3.3.3).

It has been applied to bibliometrics [3–9], question-answering systems [10], blogs [11, 12], news [13], weeding in libraries [14], and editorial strategies[15]. Many reports [1, 16–19] have been formulated to support and provide recommendations to apply ML in libraries and museums. In order to responsibly apply ML in libraries, organizations have to critically engage, manage bias, and mitigate potential harm for a more equitable and just vision of ML for the communities they serve (refer to Sect. 11.3.3 for more details). Computational systems are loaded with their creators' biases and oversights. Some examples of ML biases in libraries and their solutions are as follows:

- Libraries often reflect biases from dominant groups of patrons for tasks such as digitization, and because of which libraries' digital collection should not be believed to be representative or comprehensive of the library collection. The problem of representation in the library collection is older than in ML or digitization, and it is the responsibility of the librarians to rectify such biases and not reproduce them with a new scale of power.
- Limitations, biases, and oversights of library collections may produce flawed research that might not represent its community. They can result in outdated or offensive subject terms for a project that uses library data to categorize data based on subject terms. Such terms can be found as tags in ImageNet library,[19] which is considered a standard training dataset for image-based ML projects. Thus, catalogers should update subject terms in metadata systems frequently, and researchers should be cautious about the legal and ethical issues related to ML.
- Data teams composed of small groups of people result in gender and racial biases when creating or updating an information system as they are not representative of the country's population. These groups choose which resources should remain in the archives and which resources should be selected for digitization and computation (including ML). For instance, the study by Fagan [20] showed that the data was skewed to the dominant white population newspapers and excluded black and other minority-run papers due to the prioritization of geographic representation for state-level digitization in the Library of Congress's Chronicling America[20] project. They selected one newspaper from each state's most populous city, thus creating biasness related to racial and cultural representation. The current state of a dataset is an unavoidable reality, and libraries should work on managing bias instead of eliminating them as it is not a feasible goal for ML

---

[17] https://about.proquest.com/products-services/TDM-Studio.html.

[18] https://constellate.org/.

[19] https://www.image-net.org/.

[20] https://chroniclingamerica.loc.gov/.

projects. Hence, libraries should construct unbiased datasets and should work towards equity.

- Many ML projects seek to understand patrons, such as performing sentiment analysis on visitors' surveys and comments, analyzing patrons' borrowing habits for planning future collection development, or employing facial recognition to identify patrons using library resources or checking out materials. Studying such activities requires *participant's content*. The participants should be well-informed about the process, such as the source of data; the methods which will be used to transform, weigh, and evaluate the data; and the procedure that will result in the final decision. Further, ML projects that affect the patrons directly should be advertised over all media channels and should be thoroughly documented. The patrons should be provided with a choice to decide if they want to participate in the ML project or not. The project should be regularly audited and re-evaluated by the affected community and the library staff.

8.2.1   Challenges

8.2.1.1   Machine-Actionable Data

The collection digitized by libraries is influenced by racist, hetero-normative, gender-normative, colonialist, and supremacist biases. Libraries need to prioritize more inclusive and comprehensive digitization for ML projects. If inclusive datasets are not present to researchers, then the ML projects will reflect the biases and gaps present in the datasets.

8.2.1.2   Training Datasets

There is a lack of domain-specific training data, which presents a vital hurdle to apply ML in libraries. Further, the metadata and bibliographic information already present in libraries are not enough for the annotations needed in ML projects. Neither researchers nor librarians should assume that the existing datasets are straightaway ready for ML tasks.

8.2.1.3   Limitations of Standard Datasets

Database such as ImageNet[21] has become a universal training data for ML projects. However, it has many racial, misogynistic, and gender biases in portraying the images of the people, such as it only consists of "male" and "female" categories

---

[21] https://www.image-net.org/.

under gender and does not include other gender identities. Thus, ML projects should not use the "standard" or already existing training or ground-truth dataset just because they are readily available or easy to use. It is essential to understand the history of the existing training data in order to recognize and address the biases it might contain.

8.2.1.4   Fitting Across Domains

Most of the datasets in ML consist of selection biases that can hinder the application to other collections across different domains or periods and might not be well-fitted for the inclusion of all the communities, regions, cultures, or periods.

8.2.1.5   New Objects

During ML analyses, new objects are created. Librarians need to think about how they should document these new objects and relate them to the original resources they support.

8.2.1.6   Sharing

Currently, there is no standard for sharing ML data. For effective sharing of ML data, the data should be adapted to legal metadata systems and vendor-controlled solutions.

8.2.1.7   Staff Expertise

For effective ML application in libraries, we need to have staff having expertise in ML. With only limited LIS programs teaching ML skills, libraries may have to use vendor solutions to access ML results as there are not enough entry-level ML-trained librarians. Thus, libraries need to build cross-sectional teams consisting of domain experts, collection stewards, project managers, data scientists, and senior-level managers who can successfully implement ML projects in libraries.

8.2.1.8   Integrating Machine Learning Results into Library Systems

Many ML projects promise to aid browsing, enhance discoverability, and enrich metadata of library collections, though, presently, the results derived from ML research are seldom integrated into library interfaces and are theoretical than practical. People are often anxious about the reliability of ML-derived results. This raises the questions of (i) how one can assess the reliability of results derived from

ML projects and (ii) how one can integrate probabilistic ML-derived results into library systems with human-created information such as hand-assigned categories, tags, summaries, etc. To bridge this gap, there should be a way that humans can interact with the algorithmic processes, so that librarians can be confident about providing ML-derived results to their patrons. Moreover, as most of the libraries are depending on vendors for cataloging systems, this make ML-derived data integration more challenging as libraries do not directly maintain the systems.

Chreston Miller and **Michael Stamper** are the University Librarians at the Virginia Polytechnic Institute and State University. They performed the following study in association with **Jacob Lahne** (Assistant Professor) and **Leah Hamilton** (PhD student) from the Department of Food Science and Technology of the university.

Story: Natural Language Processing (NLP) for Sensory Science

The goal of our project, Seeing Flavors, is to take open-ended food descriptions and extract a descriptive sensory language through the application of natural language processing (NLP) to published, free-text descriptions. A set of whiskey descriptions is used as a case study. This project is a collaborative effort between the Department of Food Science and Technology and University Libraries at Virginia Tech. The novelty of the project is the application of NLP, which has been minimally used in the sensory evaluation of food. Our goal is to text mine whiskey reviews in order to identify key descriptors of whiskey. In essence, these descriptors are a language used to describe the flavor nuances of whiskies. There are two NLP parts to the project. The first is an interactive tagging visualization that takes thousands of whiskey reviews and allows for researchers to quickly identify words that are descriptors and words that are not. This is crucial as it creates an essential training dataset for the second part, which is a deep learning model that will be used to identify these descriptors automatically. After training, this model is able to identify accurately which words in a given sentence are descriptors. The current implementation is a working prototype that achieves high accuracy (>98%).

The NLP techniques used are two-fold. First, using a frequency measure for the interactive tagger, the most frequent words are presented at the top of the list of candidate descriptors using a word cloud-type visualization. This tagger uses spaCy in Python for pre-processing the data and JavaScript for the tagger interface within an Internet browser. Second, a deep learning structure with a long short-term memory (LSTM) architecture was developed using the Python library Keras.

This project allows users to quickly and accurately identify descriptors from free form text. In a real-world case example, it can allow the user to

(continued)

perform descriptive analysis (DA), a process in Food Science, to accurately describe the products. DA is traditionally performed by humans causing it to be time-consuming and expensive. Our approach can also be used for cluster analysis for different kinds of whiskies. Currently, the tagger program runs in a web browser for easy access, and we have a Google Colaboratory notebook that can run the deep learning prototype. Once set up, all the user needs to do is write in a text box, and the deep learning prototype highlights what it predicts are descriptors.

We did run into a few challenges along the way. The first was a need to tag which words are and are not descriptors. Our solution was the interactive tagger program. The other challenge was putting the pieces together for the deep learning model framework. The core code was provided through a blog post, and we needed to understand it very well to use it properly. The end product is a well-defined pipeline for data processing and descriptor identification, which is still maintained by the team, as the project is in the prototype phase.

Supplementary Media Links

1. https://lib.vt.edu/magazine/spring-2020/data/fluent-in-flavor.html
2. https://video.vt.edu/media/t/1_gh1no45n
3. https://icat.vt.edu/projects/2019-2020/major/seeing-flavors.html
4. https://vtnews.vt.edu/articles/2020/03/univlib-whiskeydescriptor.html
5. https://theroanokestar.com/2020/06/10/tech-researchers-use-machine-learning-to-build-flavor-language-for-whiskey/
6. https://www.thedrinksbusiness.com/2020/06/scientists-use-ai-to-standardise-whiskey-tasting-notes/

Lighton Phiri is a Researcher at the University of Zambia, Zambia.

Story: *The University of Zambia (UNZA) has had a functional Institutional Repository (IR) for way over a decade now. One of the key digital objects ingested into the IR is electronic theses and dissertations (ETDs). While there is a reasonable uptake of ETDs, because of the flawed nature of the ingestion process, the metadata associated with ETDs generally has missing key metadata elements and is mostly inaccurate. We have been working towards implementing machine learning models to aid in the automatic re-classification of ingested ETDs and, more importantly, enable the implementation of tools that will aid librarians at UNZA to classify ETDs correctly. The work aligned with this was presented at ETD 2019, and, recently, a submission was accepted to be published by the International Journal of Metadata, Semantics and Ontologies (IJMSO).*

8.2.2   Use Cases

8.2.2.1   Optical Character Recognition (OCR)

OCR is a supervised ML method where a model trained on example documents transcribes an unknown set of documents. It has been used by HathiTrust,[22] EEBO,[23] ECCO,[24] Chronicling America,[25] and many more large- and small-scaled digitization projects. It has significantly increased the accessibility of text archives.

8.2.2.2   Handwriting Recognition

Like OCR, handwriting recognition is a supervised ML method used to transcribe library manuscript collections, including diaries, letters, papers, and other hand-written genres. A considerable amount of training data is generally needed to train a model for a manuscript collection. However, the limited number of manuscripts for any collection in a library makes it difficult to train the model. Handwriting recognition helps to overcome the above problem by providing training data where volunteers transcribe primary texts (whether handwritten or printed), which is then used to train the model and automatically transcribe more images from the same domain, for instance, the Transkribus project by the University of Innsbruck.[26]

8.2.2.3   Crowdsourcing

Library crowdsourcing projects involve annotating collections that are widely used by patrons and are typically used as a training data or for other ML purposes, for instance, *Beyond Words* project by the Library of Congress where the volunteers improved the image metadata for the newspaper corpus of Chronicling America and used it for training ML projects like IDA[27] and Newspaper Navigator,[28] and Smithsonian Institution's crowdsourcing project [21] where volunteers transcribed digitized bee specimens' labels to train deep learning models by classifying the species. With the help of volunteers, library crowdsourcing projects can quickly build sufficient training data for domains that have little or no training data existed. Crowdsourcing is a patron-focused ML application and should require

---

[22] https://www.hathitrust.org/.

[23] https://about.proquest.com/products-services/databases/eebo.html.

[24] https://www.gale.com/primary-sources/eighteenth-century-collections-online.

[25] https://chroniclingamerica.loc.gov/.

[26] https://readcoop.eu/transkribus/?sc=Transkribus#archive-content.

[27] https://labs.loc.gov/static/labs/work/experiments/final-report-revised_june-2020.pdf.

[28] https://dl.acm.org/doi/pdf/10.1145/3340531.3412767.

the patron's consent with a detailed explanation about the crowdsourcing project's goals, management, workflow, and data models.

8.2.2.4 Discoverability

ML aids in discovering large-scale collections in libraries such as digitizing documents; extracting metadata from the digitized collection; identifying common topics across documents; grouping similar resources, be it textual, visual, auditory, or geographical; and tagging the content of digital collections. For instance, *Ex Libris's Data Analysis Recommendation Assistant (DARA)*[29] provides recommendations to improve user's experience with the database, the *Neural Neighbors*[30] project used convolution neural networks to relate resources from the National Gallery of Art to determine the similarity from Meserve-Kunhardt collection for more than 27,000 historical photographs, and the *SherlockNet*[31] project of British Library used convolution neural networks to tag over 1 million book illustrations. Thus, whether the data is textual, tabular, visual, or auditory, ML classification and clustering methods complement human cataloging and make the collection much more tractable through search and browsing.

8.2.2.5 Metadata Recognition and Extraction

ML can identify specific and meaningful metadata from a large corpus of text using the named-entity method, such as identifying geographical features across different resources and creating a geographical knowledge representation system [22]. Such projects attempt to recognize and document various metadata types to enrich library catalogs and can be integrated into library interfaces for discovery that will result in rich, contextual, and deeply linked ML-driven catalogs for library collections.

8.2.2.6 Library Administration

ML can help to perform various administrative tasks such as collection management and user behavior. Evaluation and weeding collections are vital tasks of collection management that can be performed using ML. Wagstaff and Liu [14] performed a study to train a model for weeding based on librarians' priorities and received statistically significant results between the classifier's predictions and human decision.

---

[29] https://knowledge.exlibrisgroup.com/Alma/Product_Documentation/010Alma_Online_Help_(English)/050Administration/\DARA_%E2%80%93_Data_Analysis_Recommendation_Assistant.

[30] https://dhlab.yale.edu/neural-neighbors/.

[31] https://data.bl.uk/research/sherlocknet1.html.

Preservation and conservation is also an important library task that can be performed using ML. Digitization projects can review library resources for noticeable signs of wear or damage that may need direct attention.

8.2.2.7 Chatbots

Chatbots use AI-assisted reference systems to interact with patrons utilizing the corpus of questions archived in ask-chat and ask-us systems where a subject librarian reviews the questions for accuracy, completeness, and bias mitigation [23].

8.2.2.8 Knowledge Organization

ML methods can automatically group articles and assign class numbers using classification systems to assist in cataloging [24].

8.2.2.9 Linked Data

ML methods help to identify links between resources by automatically mapping metadata that humans may not notice when browsing the Online Public Access Catalog (OPAC). By linking related resources in a digital library, ML helps researchers and librarians to explore domains effectively and efficiently. For instance, the University of North Carolina at Chapel Hill Libraries used ML to develop more comprehensive reviews [18].

8.3   Case Study: Predictive Modeling of Documents Using RapidMiner

8A

About the Case Study The case study has been adapted from Lamba and Madhusudhan [4] to illustrate a predictive modeling problem using RapidMiner and provides a step-by-step tutorial to solve it.

Problem If you have tagged documents and want to classify untagged documents of similar type in the same subject automatically.

Table 8.1 Summary of labeled topics from Case Study 4.4(4B)

Label	Tagged topic
Topic a	Children literature
Topic b	Academic library
Topic c	Information retrieval
Topic d	Archival science
Topic e	User study
Topic f	Digital library
Topic g	Library leadership
Topic h	Digital communication

Dataset Eight folders contain plain text files[32] of 441 electronic theses and dissertations (ETDs) retrieved from ProQuest Dissertations and Theses (PQDT) Global database from 2014 to 2018 for the library science subject. Each text file is segregated into different folders based on their topic proportion after topic modeling (refer to Case Study 4.4(4B) for more details). Table 8.1 summarized the labeled topics for each folder which were a result of Case Study 4.4(4B).

About the Tool   Refer to Chap. 10, Sect. 10.2.3, to know more about `RapidMiner`.

Theory Predictive modeling is the process of modeling an appropriate model according to a corpus of data for future automated classification without repeating the same procedure again and again. Support vector machine and naïve Bayes are some of the best algorithms which fit the textual data. Support vector machine (SVM) is a classification and regression algorithm that optimizes and maximizes the "predictive accuracy while automatically avoiding over-fitting the training data. It projects the input data into kernel space and then builds a linear model in that kernel space. It performs well in applications such as classifying text, recognizing hand-written characters, classifying images, as well as bioinformatics and biosequence analysis. It is the standard tools for ML and data mining" [4].

Methodology Figure 8.9 shows the workflow used to perform the topic modeling of documents in RapidMiner.

The following screenshots demonstrate the steps which were taken to perform predictive modeling using the `RapidMiner` platform:

Step 1: The folders for each topic were uploaded using the "Process Documents from Files" operator followed by text pre-processing of the text files, which included tokenization, stemming, transformation of cases to lowercase, filtering of stopwords, and generation on n-gram tasks (Fig. 8.10).

Step 2: The dataset was randomly split into 70:30 ratio where 70% (308) of data was used to train the model and 30% (133) of the data was used to test the model using the split-validation technique, and the SVM was the classifier (Fig. 8.11).

---

[32] https://github.com/textmining-infopros/chapter8/tree/master/8a_dataset.

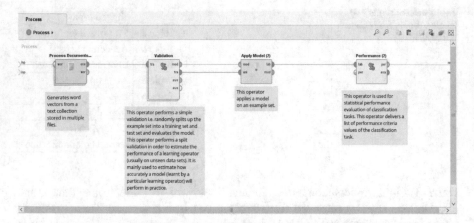

Fig. 8.9 Screenshot showing the workflow for predictive modeling in RapidMiner

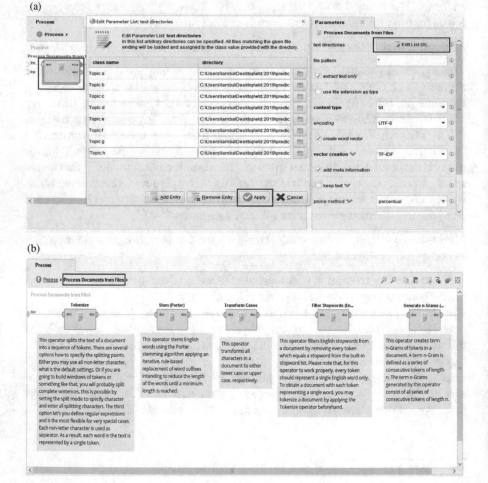

Fig. 8.10 Screenshot showing step 1. (**a**) Uploading text folders. (**b**) Text pre-processing

(a)

(b)

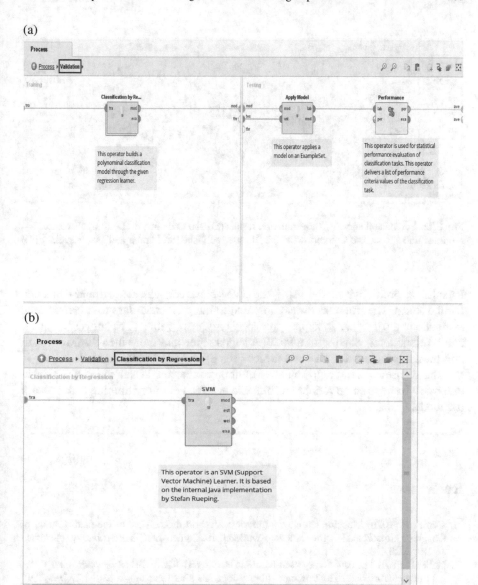

Fig. 8.11 Screenshot showing step 2. (**a**) Sub-process of validation operator. (**b**) Sub-process including SVM classifier

kappa: 0.997

	true Topic a	true Topic b	true Topic c	true Topic d	true Topic e	true Topic f	true Topic g	true Topic h	class precis...
pred. Topic a	50	0	0	0	0	0	0	0	100.00%
pred. Topic b	0	111	0	0	0	0	0	0	100.00%
pred. Topic c	0	0	45	0	1	0	0	0	97.83%
pred. Topic d	0	0	0	41	0	0	0	0	100.00%
pred. Topic e	0	0	0	0	58	0	0	0	100.00%
pred. Topic f	0	0	0	0	0	29	0	0	100.00%
pred. Topic g	0	0	0	0	0	0	65	0	100.00%
pred. Topic h	0	0	0	0	0	0	0	41	100.00%
class recall	100.00%	100.00%	100.00%	100.00%	98.31%	100.00%	100.00%	100.00%	

Fig. 8.12 Screenshot showing the evaluation results (©2020 Cadernos BAD, all rights reserved—reprinted under Creative Commons CC BY license, published in Lamba and Madhusudhan [4])

Results A predictive model using the SVM classifier was created and evaluated for the study. The true class was compared to the predicted class to determine the evaluation metrics, that is, kappa, precision, and recall values. The suggested model can be considered good with a 99.70% kappa value and more than 90% for recall and precision values for all eight topics (Fig. 8.12). The limitation of this study is that this dataset was not representative of library science ETDs in the PQDT Global database. To have some reliable results, we need to add more training data to learn the model.

References

1. Cordell R (2020) Machine learning + libraries: A report on the state of the field. Library of Congress. https://labs.loc.gov/static/labs/work/reports/Cordell-LOC-ML-report.pdf. Accessed 18 Dec 2020
2. Miller J (2020) The new library user: Machine learning. EDUCAUSE Review 55(1). https://er.educause.edu/articles/2020/2/the-new-library-user-machine-learning. Accessed 27 Dec 2020
3. Shmueli G (2010) To explain or to predict? Statistical Science 25:289–310. https://doi.org/10.1214/10-STS330
4. Lamba M, Madhusudhan M (2020) Mapping of ETDs in ProQuest dissertations and theses (PQDT) Global database (2014–2018). Cadernos BAD 2019:169–182. https://www.bad.pt/publicacoes/index.php/cadernos/article/view/2034/pdf
5. Lamba M, Madhusudhan M (2018) Metadata tagging of library and information science theses: Shodhganga (2013–2017). In: ETD 2018 Taiwan beyond the boundaries of rims and oceans: Globalizing knowledge with ETDs, Taipei, Taiwan. https://zenodo.org/record/1475795#.X9zRTdgzYuE
6. Lamba M (2019) Text analysis of ETDs in ProQuest dissertations and theses (PQDT) global (2016–2018). In: ICDL2019 digital transformation for an agile environment, New Delhi, India, pp 734–745. https://zenodo.org/record/3545907#.X9zRJtgzYuF

7. Lamba M, Madhusudhan M (2019) Metadata tagging and prediction modeling: Case study of DESIDOC. J Libr Inf Technol (2008–2017) World Digital Libr 12(1):33–89. https://doi.org/10.18329/09757597/2019/12103

8. Mahdi AE, Joorabchi A (2011) Automatic subject classification of scientific literature using citation metadata. In: Ariwa E, El-Qawasmeh E (eds) Digital enterprise and information systems. Springer, Berlin, Heidelberg, pp 545–559

9. Kossmeier M, Heinze G (2019) Predicting future citation counts of scientific manuscripts submitted for publication: A cohort study in transplantology. Transplant International 32:6–15. https://doi.org/10.1111/tri.13292

10. Momtazi S (2018) Unsupervised latent dirichlet allocation for supervised question classification. Inf Process Manag 54:380–393. https://doi.org/10.1016/j.ipm.2018.01.001

11. Chen L-C (2017) An effective LDA-based time topic model to improve blog search performance. Inf Process Manag 53:1299–1319. https://doi.org/10.1016/j.ipm.2017.08.001

12. Hasanain M, Elsayed T (2017) Query performance prediction for microblog search. Inf Process Manag 53:1320–1341. https://doi.org/10.1016/j.ipm.2017.08.002

13. Nelson LK, Burk D, Knudsen M, McCall L (2018) The future of coding: A comparison of hand-coding and three types of computer-assisted text analysis methods. Sociological Methods Res 50(1):3–44. https://doi.org/10.1177/0049124118769114

14. Wagstaff KL, Liu GZ (2018) Automated classification to improve the efficiency of weeding library collections. J Acad Librarianship 44:238–247. https://doi.org/10.1016/j.acalib.2018.02.001

15. Mrowinski MJ, Fronczak P, Fronczak A, Ausloos M, Nedic O (2017) Artificial intelligence in peer review: How can evolutionary computation support journal editors? PLOS ONE 12:e0184711. https://doi.org/10.1371/journal.pone.0184711

16. Padilla T (2019) Responsible operations: Data science, machine learning, and AI in libraries. OCLC RESEARCH POSITION PAPER, Dublin, Ohio: OCLC Research. https://doi.org/10.25333/xk7z-9g97. Accessed 18 Dec 2020

17. Lorang E, Soh L-K, Liu Y, Pack C (2020) Digital libraries, intelligent data analytics, and augmented description: A demonstration project. Library of Congress. https://digitalcommons.unl.edu/libraryscience/396/. Accessed 21 Dec 2020

18. Murphy O, Villaespesa E (2020) AI: A museum planning toolkit. Goldsmiths, University of London. https://themuseumsainetwork.files.wordpress.com/2020/02/20190317_museums-and-ai-toolkit_rl_web.pdf. Accessed 21 Dec 2020

19. Fjeld J, Achten N, Hilligoss H, Nagy A, Srikumar M (2020) Principled artificial intelligence: Mapping consensus in ethical and rights-based approaches to principles for AI. SSRN Electron J. https://doi.org/10.2139/ssrn.3518482

20. Fagan B (2016) Chronicling white America. American Periodicals 26(1):10–13. https://dx.doi.org/10.17613/M6VD5X. Accessed 21 Dec 2020

21. Earl C, White A, Trizna M, Frandsen P, Kawahara A, Brady S, Dikow R (2019) Discovering patterns of biodiversity in insects using deep machine learning. Biodiversity Inf Sci Standards 3:e37525. https://doi.org/10.3897/biss.3.37525

22. Chen H, Smith TR, Larsgaard ML, Hill LL, Ramsey M (1997) A geographic knowledge representation system for multimedia geospatial retrieval and analysis. Int J Digit Libr 1:132–152. https://doi.org/10.1007/s007990050010

23. Leung S, Baildon M, Albaugh N (2019) Applying concepts of algorithmic justice to reference, instruction, and collections work. https://dspace.mit.edu/handle/1721.1/122343. Accessed 18 Dec 2020

24. Brygfjeld SA, Wetjen F, Walsøe (2018) A machine learning for production of dewey decimal. In: World library and information congress: 84th IFLA general conference and assembly, Kuala Lumpur, Malaysia. https://library.ifla.org/2216/1/115-brygfjeld-en.pdf. Accessed 21 Dec 2020

Additional Resources

1. Sag M (2019) The new legal landscape for text mining and machine learning. J Copyright Soc USA 66:291. https://ssrn.com/abstract=3331606
2. Barocas S, Hardt M, Narayanan A (2019) Fairness and machine learning. https://www.fairmlbook.org. Accessed 17 June 2021
3. D'Ignazio C, Klein LF (2020) Data feminism. The MIT Press, United States. https://data-feminism.mitpress.mit.edu/. Accessed 17 June 2021
4. Suresh H, Guttag JV (2020) A framework for understanding unintended consequences of machine learning. ArXiv:1901.10002 [Cs, Stat]. https://arxiv.org/abs/1901.10002 [SG20]
5. Rajkomar A, Hardt M, Howell MD, Corrado G, Chin MH (2018) Ensuring fairness in machine learning to advance health equity. Ann Internal Med 169(12):866. https://doi.org/10.7326/M18-1990 [RHH+18]
6. Oxford Centre for Evidence Based Medicine (2020) Catalogue of bias. https://catalogofbias.org
7. Kliegr T, Bahník Š, Fürnkranz J (2020) A review of possible effects of cognitive biases on interpretation of rule-based machine learning models. ArXiv:1804.02969 [Cs, Stat]. https://arxiv.org/abs/1804.02969

Chapter 9
Information Visualization

Abstract This chapter aims to build a theoretical framework for information visualization with a particular focus on libraries. It exhibits and explains fundamental graphs, advanced graphs, and text and document visualizations in detail. It enumerates various rules on visual design and use cases from libraries to understand information visualization concepts and how they can be applied in libraries comprehensively. A case study showing how to build a dashboard in R language is also included. This chapter is helpful for information professionals who (i) are new to the concept of information visualization, (ii) want to know more about information visualization, or (iii) want to visualize their data.

9.1   What Is Information Visualization?

Information visualization utilizes machines for interactive visual depiction of abstract data to enhance cognition [1]. In other words, information visualization helps to visualize data with no visible or visual representation in order to solve real-life problems. It helps to solve the problem with fewer efforts in a shorter period and more accurately and can even perform things that will be impossible without a computer and a graphical representation. With computers, a high density of data can be used efficiently for visualization and also enables interaction with the digital graphical representation. Interactivity is an essential aspect of visualization. It helps to answer multiple questions posed by the user in one single chart and provides information for each new image if available.

Visualization is a way to store information out of our minds and make it accessible through our eyes and manipulations with interactive systems. In contrast, information visualization helps to transform data into a visual representation. The basic idea of visualization is to transform data into something that enhances the understanding of the given data. It teaches how to design, evaluate, and develop interactive visualizations to help people generate insights and then communicate them to other people as effectively as possible. Information visualization is the visualization of more abstracted, non-coordinate data, for example, relationship

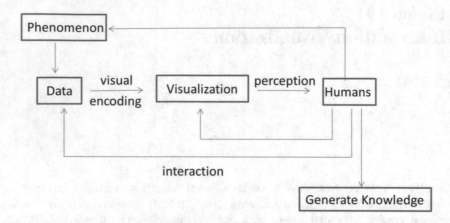

Fig. 9.1 Diagram showing the information visualization process

data, and depends on the processing of abstract data into a tangible form that an observer can effectively perceive.

Figure 9.1 shows the general information about the visualization process where data is encoded visually using a computer to create a visual representation. This visual representation is then perceived by one or more humans looking at the visual representation to get information out of it. The last step is mostly about human perception. Moreover, humans (or the users) can interact with the visualization itself and the data. The data shown in the figure is abstract. Abstract data is the data with no visible or natural visual representation and describes an object/phenomenon, which is not physical. This data visualization process helps to solve problems in a better way with less effort in a shorter time and more accurately or even performs tasks that are impossible to do without a computer or a graphical representation.

One can ask, why do we use data, or why do we visualize data? It is not because we are interested in the data itself, but we do it because it is an abstract representation of some reality or some phenomenon we find interesting. So, in general, people who are using visualization come with some pre-existing knowledge and some goal related to understanding some phenomena in a better way. Further, we want to generate some kind of useful and hopefully accurate knowledge by visualizing the data. There are many tools and software available in the market to visualize information. Some of the selected open-source information visualization tools have been covered in the next chapter.

9.1.1 Information Visualization Framework

The information visualization framework helps us to understand how data is processed for visualization (Fig. 9.2). Firstly, the data from a variety of sources

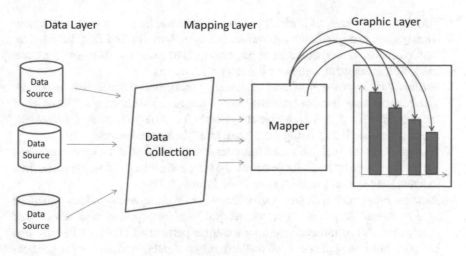

Fig. 9.2 Diagram showing the information visualization framework

are collected via data collection operation, which combines all the data from the different sources and processes them into one package for visualization. Secondly, the mapping layer maps the data and then converts it into some geometric representation. Lastly, the graphic layer displays and prepares the representation from the mapping layer to visualize communication across the visual channels for human perception.

9.1.2   Data Scale Types

The four data scale types are described as follows, where both categorical and ordinal data types are qualitative, whereas both interval and ratio data types are quantitative (Table 9.1):

Table 9.1 Data scale types

Discrete (no between values)	Continuous (values between)
Ordinal, e.g., size: S, M, L, XL **Quantitative,** e.g., numbers: 1, 2, 3 .....n	**Fields,** e.g., altitude, temperature
Nominal, e.g., shape **Categories,** e.g., nationality	**Cyclic values,** e.g., directions, hues

1. *Categorical or Nominal Scale*: It consists of variables that are non-numeric, non-overlapping, and qualitative. An arithmetic operation like addition, subtraction, multiplication, or division cannot be performed on such variables, for example, hair color, monument names, or flower species names.
2. *Ordinal or Sequence or Ordered Scale*: A quantitative scale that ranks the variables on some intrinsic ranking but not at measurable intervals. The order of the dataset is fixed, and no arithmetic operation can be performed, for example, a rating scale of 1–5 (5 being the highest and 1 being the lowest).
3. *Interval or Value Scale*: A quantitative scale where all variables have values that are fixed at some interval, for example, the temperature in Celsius, pH scale, time in hours, and spatial variables (longitude, latitude).
4. *Ratio or Proportional Scale*: A quantitative scale representing values organized in an ordered sequence with meaningful uniform spacing and has a true zero point. All arithmetic operations can be performed on this category, and comparisons can be made, such as the number of published papers per year/co-authors/citations, reaction time, and the intensity of light.

9.1.3   Graphic Variable Types

Visual elements have several properties that can be used to transmit information, and depending on the case, some of them might be more suitable than others. These properties are popularly known as graphic variables and are applied to geometric elements such as position, shape, size, hues, value, texture, and orientation. The geometric values that are used by the graphic layer to display the data for visualization in Fig. 9.2 have different geometric mappings in the perception of quantitative values to which they are mapped. Cleveland and McGill [2] sorted the graphic variables by perceptual accuracy in decreasing order as:

- Position
- Length
- Angle/slope
- Area
- Volume
- Color/density

Another study by Mackinlay [3] investigated the above order to a more exceptional level of detail and expanded that list as shown in Table 9.2. Thus, there are many different ways of connecting the data to different graphic attributes when displayed graphically for data visualization, and some of the ways mentioned in Table 9.2 are perceived better than others depending on whether the data is nominal, ordered, or quantitative.

Table 9.2 Graphic variables for data types by Mackinlay [3]

Quantitative	Ordinal	Nominal
Position	Position	Position
Length	Density	Hue
Angle	Saturation	Texture
Slope	Hue	Connection
Area	Texture	Containment
Volume	Connection	Density
Density	Containment	Saturation
Saturation	Length	Shape
Hue	Angle	Length
	Slope	Angle
	Area	Angle
	Volume	Area
		Volume

9.1.4   Types of Datasets

Every dataset can be described as a collection of items. These items have several attributes that describe the characteristics of these items and data as the collection of items and attributes (Fig. 9.3).

Items are objects/entities which we want to visualize, whereas attributes are the characteristics of the objects. Attributes are the properties of the objects/entities. Also, every object has information about attributes. Therefore, the number of items is equal to the number of attributes. There are mainly two types of datasets:

1. *Tables*: Tables are the grids or collection of rows and columns where every row represents one item or object of a dataset, and every column represents one attribute for these objects.
2. *Networks and Trees*: They are the defining characteristics of the collection of items that are linked together by some relationship. They can have attributes associated with them based on nodes (items) and edges (connected items).

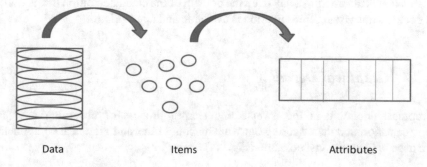

Data   Items   Attributes

Fig. 9.3 Diagram showing the components of a dataset

9.1.5 Attribute Semantics

Several predefined semantics are useful to identify the meaning of attributes, such as spatial and temporal semantics. Spatial and temporal semantics help to identify some spatial and temporal characteristics related to an attribute. Spatial characteristics include geographical region data in the form of longitude and latitude values, whereas temporal characteristics include data in the form of chronology such as date or time. Further, the essential characteristics for attribute types are to identify the following specific cases:

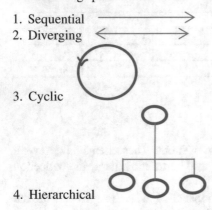

1. Sequential
2. Diverging

3. Cyclic

4. Hierarchical

9.1.6 What Is an Appropriate Visual Representation for a Given Dataset?

Data abstraction describes data to help one decide what operation and encoding methods are available and appropriate for the information. It is a way to recognize common structures in data coming from very different domains, for instance, networks showing a connection between people (friendship or relationship) or spatial distribution/phenomenon of election results by countries. Therefore, data abstraction is abstracting away from the domain of the characterization that is useful to decide what visual representation is available and appropriate.

9.1.7 Graphical Decoding

Graphical decoding is the reverse engineering process of observing a visual representation and then figuring out what mapping rules and graphical components are used to form the visualization (Fig. 9.4).

Fig. 9.4 Diagram showing
the process of graphical
encoding and decoding

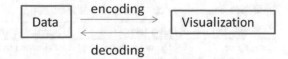

It consists of two steps: firstly, to identify the graphical components explicitly, and secondly, to identify the mapping rules. For graph visualizations, identifying visual marks like bars, lines, or points will help to determine the data items, whereas identifying the mapping rules for graphic variables like position, color, or size will help to determine the data attribute. For network and tree visualizations, the visual marks will be in the form of nodes and edges, and mapping rules will be the same as the graphs.

9.1.8 How Does One Know How Good a Visual Encoding Is?

In order to evaluate visualization and for designing or rendering visualization, there are two crucial principle guidelines to understand whether a given encoding is essential or not:

1. *Principle of Expressiveness*: It states that visual representation should represent all and only the relationships that exist in the data. Two examples where this might happen are when ordered data appears as unordered, and second is when unordered data appears as ordered.
2. *Principle of Effectiveness*: It states that the relevance of information that is being displayed should match the effectiveness of the graphic variables.

9.1.9 Main Purpose of Visualization

The three significant purposes of visualization are:

1. *Explanatory*: to explain something to someone visually.
2. *Exploratory*: to extract information from the data without knowing the content of the data. The primary purpose of this type of visualization is to help users answer questions and generate new hypotheses.
3. *Confirmatory*: to confirm the hypotheses holding the data.

The other purposes of visualization include:

1. *Analysis*: to gain insight into the data and to understand the data
2. *Communication*: to pass a message to others and make a connection with the audience/users
3. *Planning*: to predict and prepare for the future
4. *Monitoring*: to track information about performance and to be stewardship

Table 9.3 Summary of different modes of visualization

Visualization mode	User interaction	Graphics rendering	Target	Medium
Interactive visualization	Users control everything, including the dataset	Real-time rendering	Individual or collaborators	Software or Internet
Presentation visualization	Users only observe. No interaction	Precomputed rendering	Colleagues, mass audience	Slide show, video
Interactive storytelling	Users can filter or inspect details of present datasets but cannot change the data they are looking at	Real-time rendering	Mass audience	Internet or Kiosk

9.1.10   Modes of Visualization

There are mainly three modes of visualization (Table 9.3):

1. *Interactive Visualization*: It is utilized for discovery and is intended for a single individual or collaborator to control everything, including the dataset. It depends on user input and has prototype like quality, and the graphics are rendered in real time which are viewed through the medium of software or the Internet.
2. *Presentation Visualization*: It is generally used for communication and is intended for a mass group or large audience to observe. It does not support user input and is very refined. The graphics are rendered prior and are viewed through the medium of a slide show or video.
3. *Interactive Storytelling*: This visualization is somewhere between the above two and shows presentations via interactive site pages. It allows the audience to interact with the data in some restricted fashion. In this mode of visualization, the viewers cannot change the dataset, but they can investigate a little bit further, and there is more information that can be presented at once, as it would be with presentation visualization. The graphics are rendered in real time, and the target is the mass audience. It is viewed through the medium of the Internet or Kiosk.

9.1.11   Methods of Graphic Visualization

Most of the information visualization tasks need 2D computation graphics to plot and display the data (Table 9.4). Vector graphics are used to specify the 2D graphics in comparison to raster graphics, which are used to display 2D graphics. Vector graphics are mainly used for drawing, whereas raster graphics are the rectilinear array that assigns pixels.

Table 9.4 Vector graphics vs. raster graphics

Vector graphics	Raster graphics
Plotters, laser displays	TV, monitors, phones
"Clip arts," illustrations	Photographs
Postscript, PDF, SVG	GIF, JPG, etc.
Low memory (display list)	High memory (frame buffer)
Easy to draw a line	Hard to draw
Solid/gradient/texture files	Arbitrary files

9.2 Fundamental Graphs

The fundamental graphs help to visualize the combination of given attributes from the dataset. They are widely adopted, useful, and effective. They solve a considerable percentage of visualization problems. They usually consist of the x-axis, y-axis, title, annotation, data series, axis label, and legends. Sometimes a combination of two or more attributes can be shown by a single graph. Some of the famous fundamental graphs are:

1. *Bar Chart*: It helps to visualize the distribution of a quantity across a set of categories where every bar represents a category, and the length of the bar represents a quantity (Fig. 9.5). It shows the relationship between a set of categories and measured quantities that refer to these categories.

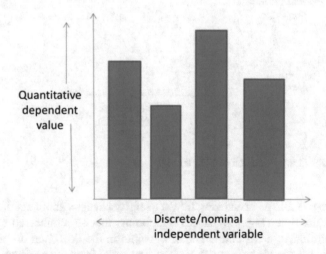

Fig. 9.5 Graphical representation of a bar chart

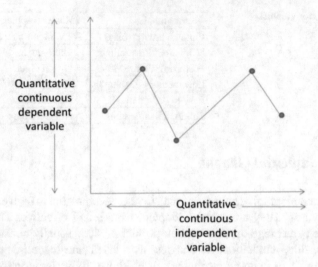

Fig. 9.6 Graphical representation of a line chart

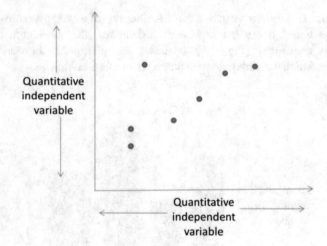

Fig. 9.7 Graphical representation of a scatter plot

2. *Line Chart*: It allows visualizing how a quantity changes about another quantity, typically the time (Fig. 9.6). It has data points that are connected by a line. It is very similar to a bar chart. The data points in the line chart are at the same altitude as the top of the bars. Therefore, it benefits from the position, but it does not have the length that one can visually see with the bars in the bar chart.
3. *Scatter Plot*: It allows us to visualize how a quantity relates to another quantity (Fig. 9.7). Every single dot represents one attribute.

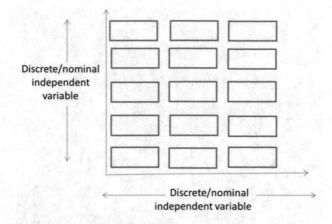

Fig. 9.8 Graphical representation of a matrix

Fig. 9.9 Graphical representation of a symbol map

4. *Matrix*: It allows visualizing the relationship between two categories (Fig. 9.8).
5. *Symbol Map*: It visualizes how a quantity distributes across two spatial coordinates (i.e., latitude and longitude) (Fig. 9.9). Every single dot/symbol represents the data point/the object for that particular location on the map.
6. *Heat Map*: These are tables where entries are displayed as the gradient of colors (Fig. 9.10). For instance, weather maps are heat maps on a table with latitude as columns and longitude as rows.
7. *Stacked Graph*: These graphs are used to visualize multiple variables simultaneously to display the change in variables along the same dimension (Fig. 9.11).

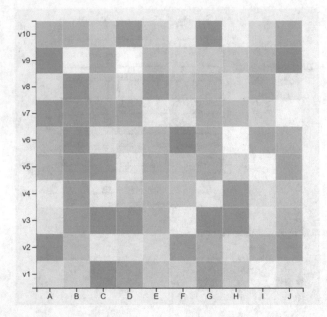

Fig. 9.10 Graphical representation of a heat map

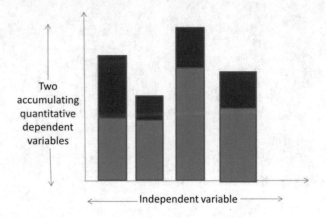

Fig. 9.11 Graphical representation of a stacked graph

9.3   Networks and Trees

The relationship between data items is beneficial when dealing with non-coordinated data or non-numerical data. Networks and trees often represent this relationship (Fig. 9.12). A network comprises nodes and edges where nodes are

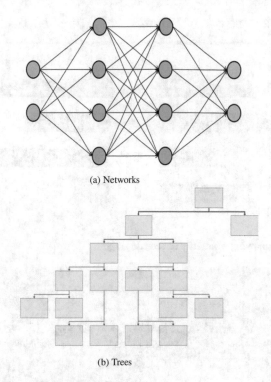

(a) Networks

(b) Trees

Fig. 9.12 Graphical representation of (**a**) networks and (**b**) trees

the data items, and edges show some relationship between two data items. Trees consist of a single-parent node that can have multiple child/siblings nodes that are minimally connected with n number of nodes and n-1 number of edges (where n can be any natural number).

9.4 Advanced Graphs

Advanced graphs are not used as much as the fundamental graphs daily. They are primarily used to solve a particular problem at hand. Some of the popular advanced graphs are:

1. *Butterfly/Tornado Graph*: It is a bar chart where two sets of datasets are displayed side by side (Fig. 9.13). It is used to compare two datasets sharing the same parameters.

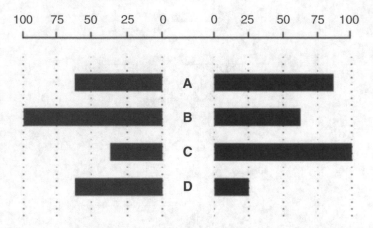

Fig. 9.13 Example showing a butterfly or tornado graph

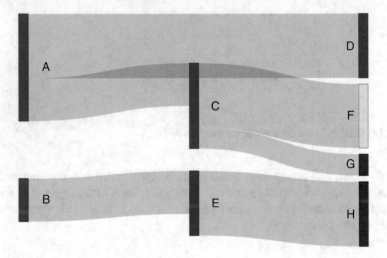

Fig. 9.14 Example showing a Sankey diagram

2. *Sankey Diagram*: It is a flow diagram in which the width is proportional to the flow rate, where there are at least two nodes (processes) to show the flow of information (Fig. 9.14).
3. *Mosaic or Mekko Chart*: This chart is used to match multiple categories or variables simultaneously (Fig. 9.15). The chart depicts the categories on one axis which are being compared, while the other axis lists the range of the dataset.

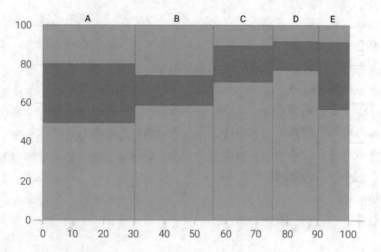

Fig. 9.15 Example showing a mosaic or Mekko chart

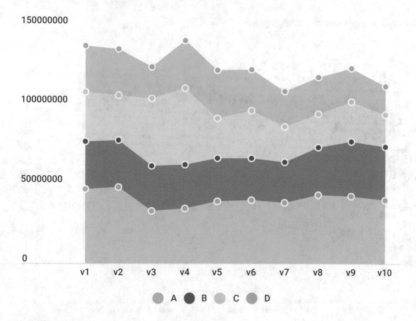

Fig. 9.16 Example showing a stacked area chart

4. *Stacked Area Chart*: It is used to visualize multiple line graphs representing a different set of categories (Fig. 9.16). The area underneath every line is commonly shaded with a selected color so that each dataset can be easily compared.
5. *Pictograph*: It uses images and symbols to illustrate data (Fig. 9.17).

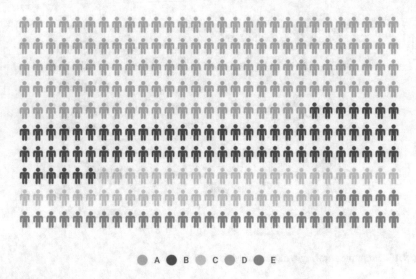

Fig. 9.17 Example showing a pictograph

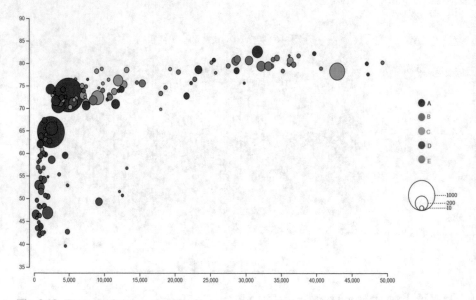

Fig. 9.18 Example showing a bubble chart

6. *Bubble Chart*: It is a multiple variable graph that is a combination of both an area chart and a scatter plot (Fig. 9.18). It utilizes a Cartesian coordinate system to plot the points on a grid with *x*- and *y*-axes as separate variables and then represents the area of the circle as the third variable.
7. *Dot Plot*: It consists of a group of data points plotted on a simple scale (Fig. 9.19).

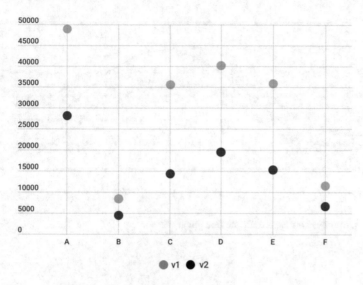

Fig. 9.19 Example showing a dot plot

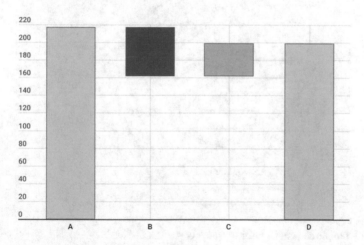

Fig. 9.20 Example showing a waterfall or Mario chart

8. *Waterfall/Mario Chart*: It illustrates the aggregated effect of consecutively introduced negative or positive values, which can either be time-based or category-based (Fig. 9.20).
9. *Multiple Line Graph*: It consists of multiple datasets with lines of varying patterns or colors (Fig. 9.21).

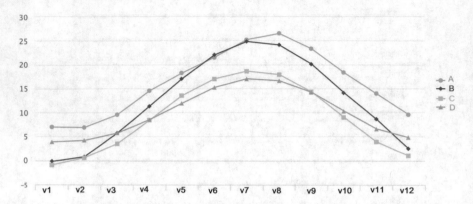

Fig. 9.21 Example showing a multiple line graph

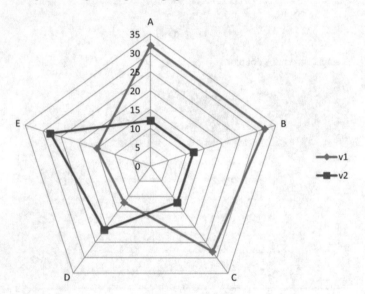

Fig. 9.22 Example showing a radar or spider or star chart

10. *Radar/Spider/Star Chart*: It displays datasets with two or more variables
    (Fig. 9.22). As each variable is charted, a line connects the point on the axis—
    making an irregular polygon that may resemble a star or spider web. It can be
    used to compare multiple datasets by representing each with a different color.
11. *Dashboard*: It provides a bird's-eye view of core visualizations about a
    particular project or process (Fig. 9.23). It is often linked to a database and
    is generally used in business analytics.

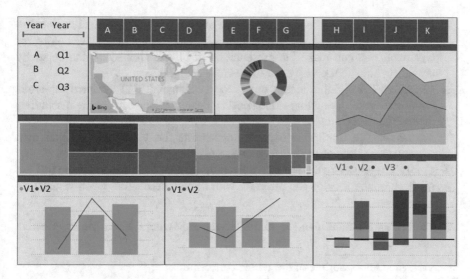

Fig. 9.23 Example showing a dashboard

9.5   Rules on Visual Design

The following rules are based loosely on Sealth Reinhold's Summary of Edward Tufte's Books [4]:

1. *Let the data speak*: The first rule says that sometimes it is best to just display the data as it is without smoothing of the data to fill in the missing data gaps and to eliminate outliers, etc., during the data processing/collection step and let the viewer take care of those details cognitively.
2. *A picture is worth a thousand words*: The second rule says that a diagram or a picture can often describe much information and can save space, which may be used with too much text or annotations to explain the same diagram or picture.
3. *Annotation*: The third rule says that the relationship between words and geometry often accomplishes the task of good visualization.
4. *Chartjunk*: The fourth rule says that adding many elements to visualization often makes it less effective at communicating its data as those elements can divert the observer from the main message that the visualization is trying to communicate. Further, it also tells us to avoid a lot of extraneous details to visualization as it may derail the observer from the actual message.
5. *The data-ink ratio*: The fifth rule is Tufte's minimalism approach, which suggests to show as much data as possible using as little ink as possible.
6. *Micro/macro*: The sixth rule allows us to focus on the various micro and macro levels of visualization by zooming in or out.

7. *Information layers*: The seventh rule makes one think of information as being organized into layers, and by using different design elements, colors, or typefaces, one can make the presentation of information somewhat better by providing hints to the viewer of the various categories.
8. *Multiples*: The eighth rule tells us to keep a constant design and have numerous visualizations utilizing that same constant design so one can see the dissimilarities as the data changes.
9. *Narrative*: The ninth rule suggests presenting the visualization either in a presentation mode or in an interactive storytelling mode.
10. *Color*: The tenth and the last rule tells us to use the colors carefully and that the rainbow gradient color can interfere with human perception when visualizing 3D shapes.

Moreover, Ben Shneiderman [5] suggested the following mantra for an effective user interface:

1. *Overview*: Get an outline of the whole collection.
2. *Zoom*: Zoom in onto the items of intrigue.
3. *Filter*: Filter out items of not interest.
4. *Details-on-demand*: Choose a group of items or a specific item and get details when required.
5. *Relate*: Visualize relationship among different items.
6. *History*: Keep a history of activities to allow replay, undo, and continuing refining.
7. *Extract*: Permit removal of subcollections or query conditions.

9.6   Text Visualization

Text visualization is a sub-domain of information visualization and is one of the essential text mining tools due to its readability by both machines and humans. As the sources of digital text have increased significantly over the recent decades in different formats (webpages, blogs, social media platforms, email, electronic resources), various approaches and techniques to analyze and visualize the knowledge generated from these sources have also increased in numbers. By nature, text documents are inherent to many challenges such as uncertainties intrinsic to natural language, irregularities, and high dimensionality. Thus, advanced techniques are needed to overcome these challenges in order to visualize text data. Some of the popular visualizations specific to text documents are:

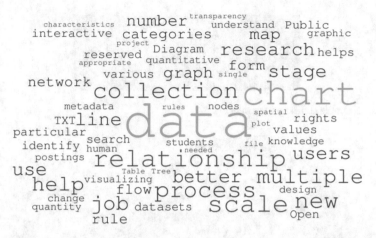

Fig. 9.24 Example showing a tag cloud

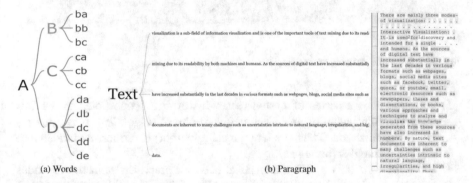

(a) Words   (b) Paragraph

Fig. 9.25 Graphical representation of a word tree. (**a**) Words. (**b**) Paragraph

1. *Tag Cloud/Word Cloud*: It is generally used to visualize relative prominent keywords or tags or terms in the text document(s) based on their frequency, weight, or rank (Fig. 9.24). These keywords are composed of single words, and their importance is characterized by different color or font size.
2. *Word Tree*: It displays text data in a hierarchical manner as tree elements using the keyword-in-context (KWIC) technique (Fig. 9.25). It usually consists of a single word connected by lines and is characterized by font size, weight, or frequency. In contrast, if a paragraph is used to make a word tree, it picks a phrase or word that presents all the various contexts where it features and arranges in a branching structure (like a tree) to uncover repetitive concepts and words.

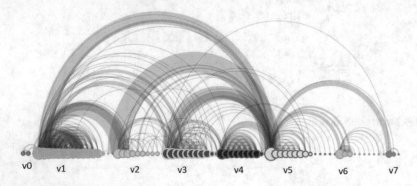

Fig. 9.26 Example showing a treemap

Fig. 9.27 Example showing an arc diagram

3. *Treemap*: It is a series of nested rectangles of size proportional to the frequency/weight value of an individual word/phrase (Fig. 9.26). It is a hierarchical visualization. Thus, the larger rectangle (root) is further subdivided into smaller rectangles (branches).

4. *Arc Diagram*: It is a one-dimensional network graph that constitutes nodes (along a single axis) and links (as arcs) (Fig. 9.27). The order of nodes is an essential strand in an arc diagram where links can be weighted.

5. *Scatter Text*: It helps to visualize two categories with their associated terms/phrases based on their frequency or weight (Fig. 9.28).

6. *Concordance Table*: It queries exact word/phrase in a text and displays the context in which that word/phrase was used (Fig. 9.29).

7. *Chromogram/Dispersion Plot*: It is used to visualize long sequences of texts by mapping the initial string of letters of a word or numbers with colors (Fig. 9.30). This visualization can reveal surprising and subtle patterns in real-world data.

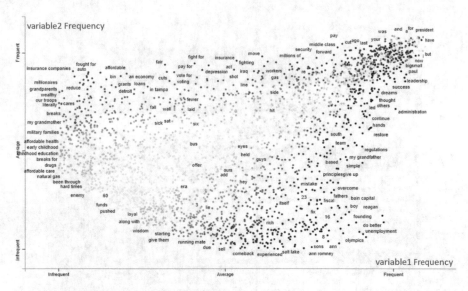

Fig. 9.28   Example showing a scatter text plot

Document	Left	Term ↑	Right
1)	other, is vastly …	live	in Perthshire, You in Sussex
1)	He was not exp…	live	many Hours, he died the
1)	She is so stupid! I	live	in the hope of seeing
1)	it may, he did not	live	for ever, but falling ill
1)	never met, we …	live	together. We wrote to one
1)	Scudamores. H…	live	within a mile of the
1)	in love that we …	live	asunder. Oh! my dear Musgrove
1)	crumbles to du…	live	an example of Felicity in
2)	that perhaps sh…	live	long. Miss Mainwaring is just
2)	This event, if hi…	live	with you, it may be
3)	he might reaso…	live	many years, and by living
3)	do not, they ma…	live	very comfortably together on the
3)	then, if Mrs. Da…	live	fifteen years we shall be
3)	if you observe, …	live	for ever when there is

Fig. 9.29   Example showing a concordance table

(a) Occurrence of words in a corpus

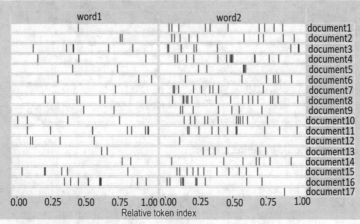

(b) Occurrence of words in different documents

Fig. 9.30 Graphical representation of a chromogram. (**a**) Occurrence of words in a corpus. (**b**) Occurrence of words in different documents

8. *TileBars/MicroSearch*: It is a visualization interface used in conjunction with a search feature and query terms (Fig. 9.31). It allows users to make one or more queries and then displays the relevancy of each query in one or more corpus. The occurrence of search terms is located as colored blocks/bars where the brightness of the colored block/bar is proportional to the relative frequency of the queries (multiple search terms are merged) within the corpora.

9. *Bubblelines*: It visualizes corpora in horizontal lines and divides segments into equal lengths where the selected words are represented as bubbles (Fig. 9.32). The size of the bubbles is relative to the distribution of the words in the corresponding corpora.

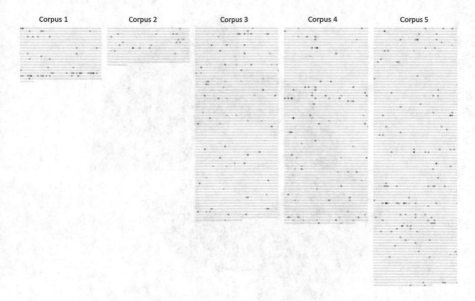

Fig. 9.31 Example showing TileBars

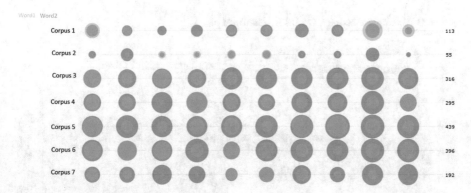

Fig. 9.32 Example showing a bubblelines plot

10. *DocuBurst/SunBurst*: It is a hierarchical visualization that spans outwards radially from parent roots to children leaves, where the roots start from the center and branches are added to the outer rings (Fig. 9.33).

11. *Streamgraph/ThemeRiver*: It is a kind of stacked area chart showing the evolution or thematic changes of groups/keywords in a collection of documents with distinct colors over time (Fig. 9.34). The graph is represented around a central axis where the x-axis presents the time and the y-axis presents the frequency/weight of a theme. It can be used to study overall relative proportions but cannot determine the evolution for individual group/entity/keyword/phrase. It is depicted to have continuity from discrete points. Thus, each theme

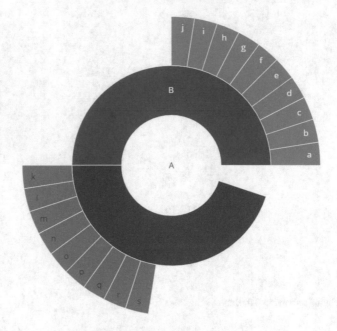

Fig. 9.33 Example showing a DocuBurst/SunBurst plot

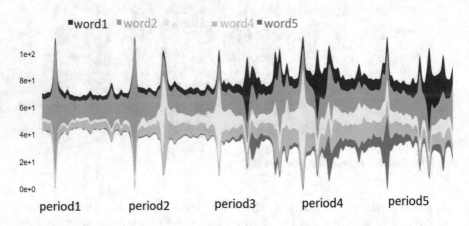

Fig. 9.34 Example showing a streamgraph

maintains the integrity as a single entity throughout the graph. Therefore, it identifies trends and patterns within a large corpus of text documents and finds occurrences or non-occurrences of themes or topics.

9.7 Document Visualization

A document can be any textual record in either physical or digital format and is frequently minimally structured and is rich with metadata (i.e., author, title, comments, date, publisher) and attributes. "Document visualization is an information visualization technique that converts textual data such as words, sentences, documents, and their relationships into a visual form" [6]. It enables users to have a better understanding of textual documents when facing a large number of text documents. It is more concentrated on visualizing documents that include metadata and attributes than text visualization that intends to visualize information with core textual content. It helps to visualize tasks such as [6]:

- Word distribution or frequency
- Semantic repetition and content
- Topics or cluster of topics in a document
- Core content of documents
- Similarity among documents
- Connections among documents
- Change in content over time among documents
- Information diffusion or other patterns

Gan et al. [6] classified the different methods of document visualization as:

1. *Single document visualization*: It emphasizes on single document content and individual words to quickly absorb text features and core content. The visualization focuses on phrases, words, contents, and semantic relations and thus is used to visualize vocabulary-based visualization using tag clouds, TextArc, and DocuBurst, semantic-structure-based visualization using semantic graphs, and document-content-based visualization using word tree and arc diagrams.
2. *Document collection visualization*: It emphasizes on enormous document collections, concepts, and themes across the collection and their relation among each other. Based on different tasks, it is used to visualize document themes using ThemeView, ThemeScapes, TopicNets, ThemeRiver, Topic Island, document core content using Document Cards, changes over different versions using History Flow, document relationships using Jigsaw, and document similarity using self-organizing map (SOM).

3. *Extended document visualization*: It is based on comprehensive tasks and attributes (other than the document) and is related to a specific field using TileBars.

9.8 Information Visualization and Libraries

"In recent years, data have become more voluminous, versatile, digitized, and accessible; new technologies have emerged to provide advanced analytical capabilities to support knowledge discovery and decision making" [7]. The exponential growth and excessive information about digital information have become barriers to access to useful information. Information visualization (InfoViz) is an exceptional methodology that can be manipulated within digital libraries to make better use of these data, where it can serve as a service mode to organize, analyze, and reveal information effectively. Applications of information visualization in the digital library can be presented in the form of expressive and comprehensible visual representation by processing abstract data present in the library [8]. Users can obtain overviews, explore, make comparisons, and search for the trends and findings from these visualizations. Digital libraries are the collaborative space where users can not only get access to appropriate resources but also receive new information such as by adding modifications, comments, and creating new relations [9]. Information visualization can be used to visualize and explore these relations between humans and human activities and the relation between users and content.

Further, information visualization can be used to evaluate text images, tags, metadata, or any other digital information through graphical means to better understand how the materials have been generated to present, communicate, and deliver messages. Following are some of the applications of information visualization in libraries [8, 10]:

1. *Supporting research*: The researchers can gain the ability to consult and collaborate with visualization experts to develop custom applications to analyze data. It will help them to reveal different layers of information embedded within the data and explore different interests. It will aid their ability in identifying potential research questions and discovering new research directions. Information visualization will provide a significant benefit to the researchers who are investigating topics using a large scale of data.

2. *Understanding library data*: Information visualization can feature real-time visualizations generated for each hour from checked-out items including books, DVDs, or CDs; analyze the text resources by tagging each document based on particular classification criteria; analyze non-text resources to provide users with multiple viewing angles so that they can observe these angles using 3D technology; visualize information retrieval process by using the semantic relationship to

understand the search result and grasp the retrieval direction to guide the retrieval process; visualize the retrieval results to express the relationship between the retrieval content using statistical, clustering, correlation, etc., in order to reveal the hidden pattern; and conduct domain knowledge visualization using semantic link network and cognitive map. This service will help to calculate the flow of information on the collection of the library.

3. *Infographics to deliver library messages*: Infographics can promote library events and activities and deliver library messages.
4. *Storytelling*: InfoViz can be used as a powerful medium of communication in addition to being a tool for discovery and data analysis. It can offer a medium for libraries to advocate by telling their stories, engaging, and convincing the stakeholders by making a personal connection to the library data. It can further engage with the community by encouraging its patrons to share their stories and experiences.
5. *Library data assessment*: Information visualization can be used as a tool for library decision-making by visualizing the library assessment data by using metrics of ARL ranking, daily gate count, and research services.
6. *Budgeting*: InfoViz can help the libraries to make informed purchasing decisions by determining the need to purchase resources or not based on their usage pattern.
7. *Suggesting titles for weeding*: Based on circulation, searching, and reading pattern, information visualization can help the librarian to identify the resources which have to be weeded out.
8. *Analyzing citations and mapping scholarly communications*: The library can map scholarly communications and analyze citations for the faculty members and research scholars using information visualization to determine the research productivity of a particular department.

Issac Williams is a University Librarian at the University of Texas, Arlington. He caters to the needs of humanities, social sciences, STEM, and professional discipline patrons.

Story: *I am the Data Visualization Librarian at UTA, and I work with students and faculty to help them develop skills in data visualization. I am a digital humanist but have worked with students from a variety of disciplines, from literature to engineering to marketing.*

Marcella Fredriksson is a Web and Discovery Services Librarian at the University of North Carolina Wilmington.

Story: *The Facts and Planning dashboard shows various data about Randall Library to our stakeholders and peers. It is built manually in HTML every month from a variety of data sources and databases.*

(continued)

Carady DeSimone (Wayne State '20) is an emerging archival professional currently on contract with the National Parks Service. As a career-change archivist, she leverages prior experience to support and advise other new professionals through the Society of American Archivists and the Society of Florida Archivists.

Story: The Roles of Tableau and TXT in a Pre-/Post-COVID Longitudinal Inquiry of Archival Labor Trends

Most archival training programs require a hands-on practicum, particularly to ensure accreditation from ALA. Students are placed in an internship role at an institution in their community and encouraged to engage in reflective analysis of their experiences practicing archival theory hands-on. I completed Wayne State University's MLIS program off campus and virtually, prior to the COVID-19 pandemic shutting down the majority of "brick and mortar" GLAM and higher educational institutions. Many of my peers were quickly placed in university archives, with corporate partners, or in local historical agencies.

Having recently relocated, in addition to a complete career change, my local professional network was next to nil. While it was technically not my responsibility to secure a host site, I knew that my advisor would have some difficulties networking on behalf of an off-campus student, particularly for a paid internship. Knowing that I would have additional barriers to the already tight job market, I began my search for a host site. While volunteer

(continued)

work is always an option, the majority of students, particularly in the early twenty-first century need some form of financial compensation for their time. Exploiting eager, desperate students as unpaid labor is at best selfish and at worst despicably unethical. Expecting students to engage in unpaid labor ignores their fundamental human needs of safe housing, healthy nourishment, and a positive work-life balance.

It was during this job search that I began collecting data. I found the third-party posting aggregator, Archives Gig, and immediately subscribed to updates. I also subscribed to many other announcement boards, including my SIS program's listserv and the SAA (for-profit) job board. Over time I added USAJOBS, LinkedIn, and others. Although I would only apply to a small portion of the postings due to constraints of location, qualifications, and long-term potential, I saved all of the emails. The resulting dataset is not foolproof by any means, but rather an organic snowball effect leading to over a year's worth of data—and the collection is still active.

Eventually, I found my host institution and completed my practicum hours. I was thrilled. I loved it. But it was a grant-funded position which offered no follow-on employment. So, back to the job boards, I went! I had never stopped collecting data, but I had slowed down on my applications.

Eager to explore existing data and enrolled in Data Visualization with Dr. Timothy Bowman[a], I decided to perform a test run. I selected the first semester of data collection as a sample of convenience. It was the smallest folder, and the semester was ended, so I would not be adding further data. Using a Tableau student license, I knew I would be able to clean and standardize the emails. . . eventually? The problem I ran into was on how to effectively transpose the data from a folder of 288 emails into some sort of structured format—short of entering them into a form one by one.

Enter AWEF: *The COVID-19 pandemic hit in the middle of my last semester, wreaking havoc globally. Through the many closures and budget cuts, our peers needed help. Lydia Tang[b] and Jessica Chapel[c] began a grassroots, ad hoc committee to explore and develop a mutual aid program for contingent workers, who were suffering the greatest financial impacts due to closures, layoffs, and hiring freezes. With the generous support of SAA administrative staff and the SAA Foundation, the Archival Workers Emergency Fund (alternately AWEF or AWE Fund) was born.*

After AWEF was founded, funded, and began disbursing awards, this compassionate group of peers, mentors, and organizers began digging into the deeper issues surrounding contingent labor. While these conversations are ongoing, fellow AWEF Organizer Bridget Malley[d] had the brilliant idea to compare my pre-COVID data to that which could be collected post-COVID. As this would constitute a longitudinal study, we will need to continue collecting data until at least June 2021 to balance the time scale.

(continued)

Saving the job posts according to semesters allows an easy-to-follow batching system. Currently, we have only processed this first semester as a test; even with the "shortcuts" illustrated, the procedure still is time-consuming and, from time to time, mind-numbing.

The Concept: *The best thing Tableau did for me was to get me to consider how to construct and record a "flow" of data in order to ensure transparency and trustworthy data stewardship. The flow in Fig. 9.35 was created using Tableau Prep Builder (TPB) and provides a convenient way to document and annotate legitimately required modifications and/or standardizations of a dataset.*

Fig. 9.35 Sample flow from Tableau Prep Builder. © Carady DeSimone, all rights reserved—reprinted with permission from Carady DeSimone

Sample Flow: Each step represents a record of change to raw data. By iteratively linking files to document the process, we can provide annotations to maintain or ensure proper data integrity. However, the gap must still be bridged between email client and Prep Builder, which best functions by handling TXT, CSV, and XLS.

This is a key concept to explore: the relationship between text, spreadsheets, and CSV. By the way, that is **C***omma-***S***eparated* **V***alues, with a "value" equating more or less to "metadata field." Keeping each entry on one line indicates to the computer that it is a single row—and the tab/comma/semicolon serves as a marker to enter the data on a new cell (column).*

So wait, why comma? Say it with me, archives folks: "Well, it depends..." ...on your preference. In time you will come to figure out what works for your projects and what does not. Sometimes a comma is best for readability. Semicolon is good for financial data instead of a comma (which may fall within a number, depending on the formatting). Tab is my favorite as it is the most visually appealing, but it can get a little tricky if many of the columns are blank or if the number of characters varies. Data that is coded to a fixed number of characters looks very smooth with tabs inserted. See Fig. 9.36 for an example of not smooth data.

(continued)

The Process:

Stage 1: Prep for Prep
From Outlook > Print to PDF >
This step only applies to extracting data from Outlook. If you are working with data that is already in a TXT, CSV, or XLS format, you can likely skip ahead.

- *Keep files in a dedicated directory throughout data collection.*
- *Open Outlook.*
- *Navigate to the directory that holds the collection of emails.*
- *Highlight and select all of the messages in the folder:*

  - *Try Ctrl+A, but sometimes the number of items prevents this from working.*
  - *Since multiple items are selected, this may take some time to process.*

- *Print to PDF. This puts all of the text in one place (file, container).*

  - *Most computers are set up to include Microsoft or Foxit Print-to-PDF listed as a printer. If this is not an option under "Print" on your system, ask your IT folks about it.*

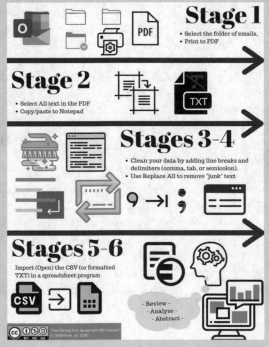

(continued)

Stage 2: Extract
Ctrl-A; Ctrl-C > Ctrl-V

- *Open the PDF you just made.*
- *Ctrl+A to highlight All.*
- *Ctrl+C to copy.*
- *Ctrl+V to paste in your desired application.*

  - *There are two options at this point: Notepad or Word (or alternative). Stripping the formatting via Notepad makes the text easier for computers to process and is faster than other methods. Experiment with both options if you are unsure!*

- *SAVE. Save frequently, save often.*

  - *I try to make a habit of saving before and after each stage of processing; usually, this would be regarding a format change (e.g., TXT to CSV, PDF to TXT).*

Stage 3: Quick Stripping
This stage also takes place in Notepad. Now we have a text-only record of the data. But still very messy. Begin cleaning your data by removing extraneous "junk text" using the "Replace All" function of Notepad by simply leaving the "Replace with" field blank. Be sure to notate a running list or table of all deletions from the text for reference in an Appendix (full disclosure). Examples of what to delete include pre-formatting or automatic text such as names, email signatures, and common headers or footers. It is not necessary to preserve the count of replacements, as not all messages will be the same; however, this iterative process will distill redundant and extraneous data from the relevant aggregated titles.

Advanced Option: Use "Replace All" to insert a filler character of your choice if it will be helpful later—for example, a particular email footer could be replaced with a standardized visual flag such as "end." This technique is also a good way to standardize common misspellings or anonymize rows of metadata.

After this stage, it is appropriate (and recommended!) to set the project aside for a couple of days to give your brain a break. Going forward, you will want to be sure to work in identifiable "chunks" or batches. Oh, and save again!

(continued)

Stage 4: Standardize

Now we will begin to standardize the data in a way that will make it computer-friendly. I have found no other alternative to human intellectual capital. However, I have cheat codes! I learned this concept from an IT mentor while working with batched financial data and institutional mailing lists. The principle is so simple that it is very easy to overlook! But you have to first know a bit about how computers do something called "parsing." In order for a computer to parse data, it must first be standardized. This stage also takes place in Notepad.

The goal is to ensure that each entry is given one line (which will become a row in the database), and the special character (tab, comma, or semicolon) will align with columns of metadata. This involves inserting a given character that will be "read" as a marker for a new cell, but that is part of the next stage.

Also, during this stage, I look for redundant entries or irrelevant email text that was not cleared up via "Replace All." This stage is most easily accomplished within Notepad, with the word wrap option off.

Fig. 9.36 Highlighting iterations of tabs in actual data. © Carady DeSimone, all rights reserved—reprinted with permission from Carady DeSimone

In this actual data file, I have highlighted the first and second tab spaces in yellow and green, respectively. It is a tedious, line-by-line labor of love, but for the right person, it can be rather zen and sometimes even fun. Just remember to be as ergonomic as comfortable, take breaks, and save often.

Stage 5: (Re)Formatting

After you have saved your shiny, standardized TXT file, you will want to open up your spreadsheet program (I use MS Excel out of habit). Shortcut: from notepad, save as CSV. Then open that file in your spreadsheet program.

- Select "Open" file.
- Navigate to your working directory, where the project "lives."

(continued)

- *You may need to change "file types" to "All Files" in the dropdown menu. This will include all files in the directory, including your TXT from earlier.*

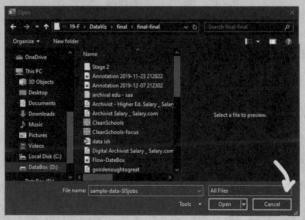

When you select the TXT file through Excel, it should automatically trigger a popup like this:

(continued)

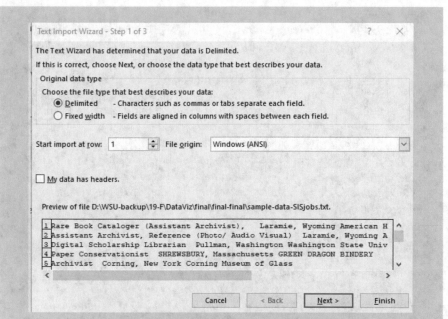

- *Verify that the data type is set to Delimited.*
- *In this case, I do not have headers, so I leave the box unchecked.*

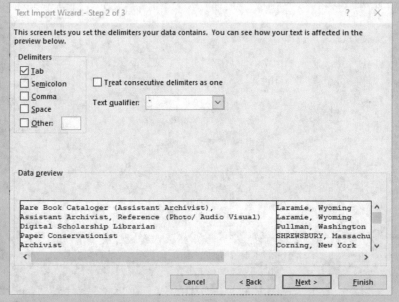

(continued)

The next step is selecting your delimiters or markers. Tab is automatically checked, so if you have used a different marker, now is the time to tell your PC! The preview will reflect the column breaks.

Now, you can either "Finish" or set data types for each column. I usually skip ahead because my goal of ingesting the data into Excel is to be able to format, filter, and further clean it. I look for discrepancies, bad formatting, and nonstandard data.

Note: You can ingest a variety of formats into Tableau Prep Builder— including PDF, TXT, CSV, XLS, JSON, and more. However, I still find myself turning back to the TXT-CSV method prior to using TPB for standardizing data. Notepad is (in my experience so far) still the easiest to visually break down aggregated data into standardized, machine-readable files.

Stage 6: Abstraction

In the cases where postings included text versions of the listing, we can further extract thematic elements of the current expectations of archive graduates and LIS professionals. In line with Johnny Soldaña's Manual for Qualitative Researchers, we can brew a "word cloud" of common skills and requirements and emphasize them according to their frequency (e.g., implied importance).

This map shows a list of ALA-accredited SIS/iSchools throughout America[e], with those offering archival programs highlighted. I overlaid these markers onto a sample of job postings from Archives Gig, an independent job aggregation site. I started collecting posts in Winter 2019 semester [Jan–Apr 2019] and have broadened the number of sources since then.

(continued)

Data Sources

Currently included datasets:

- *SAA19 Archivist Salary Transparency Open Spreadsheet (open source, 2019)*
- *Art + All Museums Salary Transparency Survey (approved, 2019)*
- *Self-collected LIS/LAM job postings (2018–2020+)*
- *Academy of Certified Archivists job postings (released)(2019)*
- *Northwest Archivists job postings (released)(2019, partial)*

Pending datasets:

- *Art + Museum Transparency End Unpaid Internships Spreadsheet*
- *AR Gov salary transparency (released, pending)*
- *AR Universities salary transparency (released, pending)*
- *LR Local gov't salary transparency (released, pending)*
- *Adjunct Salaries (open source, pending)*
- *Library Worker/Salary Info Spreadsheet (open source, pending)*
- *Society of Florida Archivists job postings*
- *Midwest Archivists job postings*

Other data sourced from:

- *CareerOneStop. (2019). Occupation Profile for Archivists. Retrieved November 30, 2019, from* https://www.careeronestop.org/Toolkit/Careers/Occupations/occupation-pro-file.aspx?keyword=Archivists&onetcode=25401100&location=UNITEDSTATES.
- *O*Net Online. (2019, November 19). Details Report for: 25-4011.00— Archivists. Retrieved November 30, 2019, from* https://www.onetonline.org/link/details/25-4011.00.
- *U.S. Bureau of Labor Statistics | Division of Occupational Employment Statistics. (2018, May). Occupational Employment Statistics Query System. Retrieved November 30, 2019, from* https://www.bls.gov/oes/.
- *US Census Bureau. (2019, September 26). 2018 Median Household Income in the United States. Retrieved November 30, 2019, from* https://www.census.gov/library/visualizations/interactive/2018-median-household-income.html.
- *World Population Review. (2019, October 8). Cost Of Living Index By State 2019. Retrieved November 30, 2019, from* http://worldpopulationreview.com/states/cost-of-living-index-by-state/.

(continued)

Sometimes, it is confusing as to what type of visualization will be appropriate for a particular type of library task. Bertini et al. [9] described the usage of a particular type of visualization appropriate for a particular type of digital library tasks (Table 9.5), whereas Hu et al. [11] proposed four visualization modules for resource structure, resource coverage, literature knowledge discovery, and reader knowledge discovery. The information from both the studies will be most beneficial for the LIS professionals and students who are new to information visualization and want to visualize data from their library.

9.8.1 Use Cases

9.8.1.1 University of North Texas Libraries

The Tableau profile[1] of the University of North Texas Libraries presents the results from various assessments such as *ebook usage,*[2] *collection usage 2020,*[3] *circulation trends,*[4] *inter-library loan requests,*[5] *e-resources,*[6] *journal usage,*[7] and *subject analysis*[8] for their university libraries' collections. The University of North

---

[1] https://public.tableau.com/profile/untlibraries#!#%2Fvizhome%2FJournalism_CollectionEvaluation2020%2FCourses.

[2] https://public.tableau.com/app/profile/untlibraries/viz/EbookUsagebyCollection/TitlesUsedbyCollection.

[3] https://public.tableau.com/app/profile/untlibraries/viz/Usage_15929267793100/AnnualCirculation.

[4] https://public.tableau.com/app/profile/untlibraries/viz/Circulations/CirculationSummary.

[5] https://public.tableau.com/app/profile/untlibraries/viz/ILLRequestsbyCollection/RequestsbyYearwTrendline.

[6] https://public.tableau.com/app/profile/untlibraries/viz/E-Resources/NumResourcesbyCollectionTreepmap.

[7] https://public.tableau.com/app/profile/untlibraries/viz/LUC-JournalUsage/JournalUsageoverTime.

[8] https://public.tableau.com/app/profile/untlibraries/viz/CollectionEvaluationSubjectAnalysis/QualitativeMatrix.

Table 9.5 Summarization of correspondence between digital library task and visualization (© 2005 Springer Nature, all rights reserved—adapted with permission from Springer Nature, published in Bertini et al. [9])

Digital library task	Task details	Visualization
Managing digital library resources	Locating resources, creating new resources, editing existing resources, depleting resources, organizing resources, archiving resources	Tree and network graphs to represent the relationships among different resources
Managing links to related/similar resources	Creating cross-reference links among similar resources, "see also" items	Network graph
Storing metadata about resources	Creator, content, technical requirements, etc.	Tree/hierarchical visualization
Checking for inconsistencies in the metadata	N.M.[a]	Multidimensional visualizations
Handling glossaries, index facilities, thesaurus, dictionaries, and classification schemes	N.M.[a]	Trees and hierarchies to represent the relationships between/among words
Retrieving content/services usage statistics	N.M.[a]	Multidimensional visualizations to visualize the website paths, flow through the site and common click streams
Providing bookmark facility	N.M.[a]	Tree visualization where bookmark features such as time could be encoded on nodes using visual attributes like color or size where two visualizations (global and individual views) could be integrated
Supporting multilingual features	N.M.[a]	Network graph
Supporting search mechanisms	Keyword search, parametric search	Fundamental graphs
Providing navigation-related functions	Browsing predefined catalogs	Tree visualization
Filtering search/browsing results	According to some particular entities such as personal profiles, etc.	Multidimensional visualizations where web search results can be organized by topics and displayed on a 2D map where each document is represented by an icon and document aspects such as the relevance of the document to the query, documents' relationship with other documents, etc., could be coded using visual attributes such as color, size, etc.

(continued)

Table 9.6 (continued)

Digital library task	Task details	Visualization
Managing profiles	Presentation of contents according to profile and profile definition, e.g., professional interests, suggestions for content based on profile, suggestions for discussion with other library members with similar interest profiles, etc.	Multidimensional visualizations
Monitoring usage and identifying common patterns of use	Information producers, information consumers (and collaborators)	Network graph can be used to visualize the interaction between different users

[a] N.M. stands for *not mentioned*

Texas Libraries shows an excellent example of how librarians can use various visualizations to analyze library data.

9.8.1.2   Woodland Middle Library

Lindsey Kimery is a Middle School Teacher Librarian at Woodland Middle School, Williamson County, Tennessee. She prepared an infographic using Piktochart using "her school's colors and included a picture of a mural in her library" [12] (Fig. 9.37). Her library's infographic helped her "visualize the overall health of her library program and set future goals" [12]. By looking at her infographic, she said she knew she needed "to create more awareness of the library's digital content. As her school is growing, and in order to keep the items/student count, she needed to buy more books the coming year" [12]. However, she knew that her district significantly cut the budget for each school, so the school had more students and less money. Thus, her infographic helped her plan ways to "show growth for next year, and it helped her dream about what she would show and share next year" [12].

9.8.1.3   Beyond Downloads Project

Beyond Downloads project [13, 14] was an "online survey hosted at the University of Tennessee and distributed by Elsevier to ascertain a complete picture of the use and value of scholarly articles (Fig. 9.38). The project resulted from an international collaboration among the University of Tennessee, CIBER Research Ltd., Project COUNTER, and Elsevier. It was sponsored by Elsevier." The project also "helped to answer the following questions: (i) What was the average number of scholarly articles down-loaded per research project?; (ii) How did the library stack up as a source of downloaded articles versus research social networks?; (iii) What were

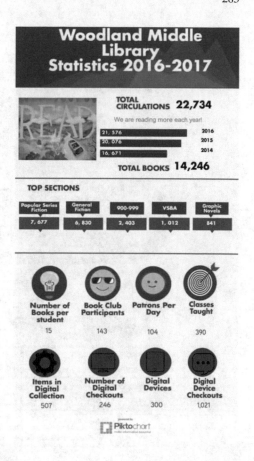

download counts missing?: (iv) How much did scholars share, and what did they
share?; (v) Were there ways to calculate or measure sharing?" [13, 14]. The project
"conducted focus groups and interviews with 29 scholars in the US and UK on how
they obtained, saved, shared and used scholarly articles. These findings informed
the development of an international survey" [13, 14]. The survey invitation was sent
by Elsevier "from roughly November 2014 to mid-January 2015 to 32,956 authors
who had published in an Elsevier journal. The survey was hosted on the University
of Tennessee website and 1,000 faculty members, researchers, and PhD/master's
students responded to 34 questions, including rating scales, demographics and open-
ended questions for commentary" [13, 14]. The survey results showed "that the
library played an important role in providing access to research and that usage
was substantially under-reported. Libraries needed to share these findings with their
administration to ensure institutional leaders understand the scholarly environment
and the role the library played in facilitating research processes" [13, 14].

Fig. 9.38 Beyond Downloads project infographic (© 2018 Elsevier, all rights reserved—reprinted with permission, published in Library Connect Newsletter [14])

9.8.1.4   Library Advocacy

"Early in 2018, then ALA President Jim Neal called for public support against (i) the closing of school libraries, (ii) the reductions in professional staffing, (iii) the erosion of budgets for resources and technology, and (iv) the consequent weakening of the librarian–teacher partnership in the classroom" [15]. The infographic in Fig. 9.39 depicts "how strategic vision and careful management had helped US public libraries weather the storm of the Great Recession, supporting their role as a lifeline to the technology resources and training essential to building digitally inclusive communities that enabled full participation in civic life and the nation's economy" [16].

Fig. 9.39 The US public
libraries weather the storm:
innovative services continue
budget cuts (© 2018
American Library
Association—reprinted with
permission by American
Library Association [16])

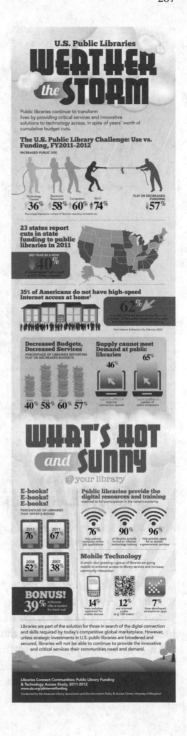

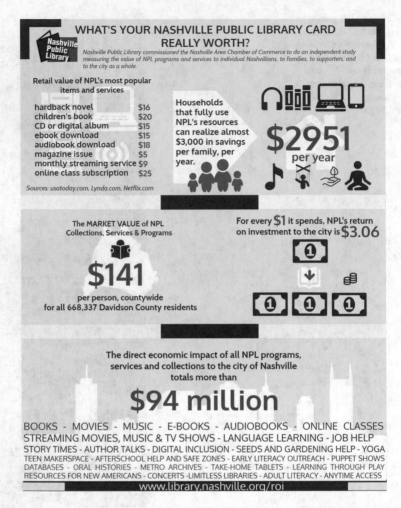

Fig. 9.40 What's your library card really worth? (© 2020 Nashville Public Library—reprinted with permission by Nashville Public Library [17])

Another infographic [17] measured how much Nashville's Public Library was worth (Fig. 9.40). It showed that the library card for Nashville Public Library was worth $2951 per year, and the library saved every resident an average of $141 a year. Further, for every $1 spent on the library by the government, the return investment to the city was $3.06. Lastly, the direct benefit of the library services and programs cost more than $94 million in impact from a $30 million budget. With unlimited activities and resources offered by the library, the library card was free for the public.

9.8.2   Information Visualization Skills for Librarians

Chen [7] emphasized that the "complexity of information visualization in this data-intensive era of digitization, understanding visualization, as well as how it can effectively be employed, is now considered as a core skill for librarians." She further accentuated that "incorporating information visualization as a component of information literacy skill for a librarian is very important, and many educational efforts in the form of MOOCs, or webinars have been created to meet the demand for greater familiarity with this new technology" [7]. Furthermore, she highlighted that "new job titles in the LIS field have emerged from the significant trend in data and visual analytics such as data and visualization librarian, and visualization and digital media librarian" [7]. Some of the vital skills that are needed by librarians to apply and use information visualization are:

1. To be acquainted with various visualization software and platforms that can be utilized to design a variety of visualizations
2. To understand the various applications of implementing information visualization in libraries to aid the needs of the users
3. To have a defined purpose of using information visualization technique in libraries
4. To contribute in the area of visual literacy by teaching the visual skills via information literacy programs to the users and staff members
5. To convey complex data using visualizations
6. To use principles of design to create visually appealing and informative visualizations

9.8.3   Conclusion

A good visualization can help to reduce *visual search time*, provide an effective apprehension of a complex datasets, reveal relationships, view dataset from several perspectives, and convey information effectively. It can create enhanced interfaces for better analysis, retrieval, management, interaction, and apprehension of resources present in digital libraries. In order to stay competitive and beneficial to the users, libraries are required to understand and invest in information visualization. The ability to process and interpret data easily via information visualization is quite significant. These progressive, forward-thinking technologies will be essential in paving the way for a stronger future for libraries. Various benefits of visualizing library resources comprise of:

1. Better comprehension of the library's resources, services, and users
2. Potential to improve library services
3. Better understanding and efficient presentation of the library's collection

4. Enhances the decision-making process regarding various administration operations
5. Simplifies the complex quantitative information
6. Identifies the areas for attention or improvement
7. Identifies the relationship between data
8. Explores new patterns and reveals hidden patterns in the data

Some of the limitations and the most common problems that are encountered
"when trying to combine information visualization tools and library data are often
the financial, technical, and human resources. Due to the lack of specialized IT skills
and the necessary knowledge to develop information visualization, most libraries
are forced to quit the idea of information visualization or rely on external IT
professionals that could increase the libraries' costs" [18]. Other challenges may
include working with users' data and intellectual property, and text resources that
are under copyright. Future trends of information visualization in libraries call for (i)
experimenting with different information graphics that use different tools and provide different results which can maximize the benefits offered by data visualization
in libraries, (ii) research on what type of information visualization techniques can
be used to communicate archival context and content in a better manner, and (iii)
investigating how information visualization intersects with information literacy.

9.9   Case Study: To Build a Dashboard Using R

9A

About the Case Study Flexdashboard is an interactive dashboard that uses an
RMarkdown file to publish a group of related visualizations. It supports a variety
of widgets and layouts. The storyboard layout presents a sequence of visualization
and related commentary. The visualizations can be viewed dynamically using the
Shiny package in R.
An RMarkdown file[9] was prepared to make a dashboard using storyboard layout
of flexdashboard in R. The dashboard can be viewed at https://textmining-infopros.
github.io/dashboard/. This dashboard summarizes the graphs/plots resulted from all
the case studies used in the book (Fig. 9.41). It consists of three essential sections:

1. *Storyboard*: It summarizes the visualization for a specific case study. For this
   case study, 12 different storyboards were prepared to summarize the results from
   all the case studies.

---

[9] https://github.com/textmining-infopros/dashboard/blob/main/flexdashboard_R.Rmd.

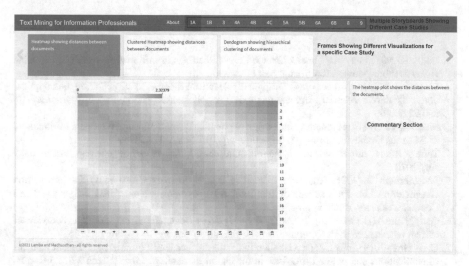

Fig. 9.41 Screenshot showing the dashboard summarizing the case studies' results from *Text Mining for Information Professionals: An Uncharted Territory* book

2. *Frame*: It shows different visualizations from a specific case study and divides them into different subsections.
3. *Commentary*: This section is used to explain the visualization.

Libraries and librarians can use storyboards to exhibit (i) their book collections in different sections, (ii) new projects or studies conducted by the faculty, and (iii) other marketing and publicity activities related to their services and products.

Virtual RStudio Server You can reproduce the analysis in the cloud without having to install any software or downloading the data. The computational environment runs using BinderHub. Use the link (https://mybinder.org/v2/gh/ textmining-infopros/dashboard/main?urlpath=rstudio) to open an interactive virtual RStudio environment for hands-on practice. In the virtual environment, open the `flexdashboard_R.Rmd` file to launch the dashboard.

Virtual Jupyter Notebook + RMarkdown You can reproduce the analysis in the cloud without having to install any software or downloading the data. The computational environment runs using BinderHub. Use the link (https://mybinder. org/v2/gh/textmining-infopros/dashboard/main?filepath=flexdashboard_R.Rmd) to open an interactive virtual Jupyter Notebook for hands-on practice.

References

1. Card SK, Mackinlay JD, Shneiderman B (1999) Readings in information visualization: using vision to think. Morgan Kaufmann, San Francisco
2. Cleveland WS, McGill R (1984) Graphical perception: theory, experimentation, and application to the development of graphical methods. J Am Stat Assoc 79:531–554. https://doi.org/10.2307/2288400
3. Mackinlay J (1986) Automating the design of graphical presentations of relational information. ACM Trans Graph 5(2):110–141. https://doi.org/10.1145/22949.22950
4. Tufte's Rules. http://www.sealthreinhold.com/school/tuftes-rules/rule_one.php. Accessed 9 Aug 2020
5. Shneiderman B (1996) The eyes have it: a task by data type taxonomy for information visualizations. In: Proceedings of the 1996 IEEE symposium on visual languages. IEEE Computer Society, Boulder, pp 336–343
6. Gan Q, Zhu M, Li M, Liang T, Cao Y, Zhou B (2014) Document visualization: an overview of current research. WIREs Comput Stat 6:19–36. https://doi.org/10.1002/wics.1285
7. Chen HM (2019) Information visualization skills for academic librarians: a content analysis of publications and online LibGuides in the digital humanities. Libr Hi Tech 37(3):591–603. https://doi.org/10.1108/LHT-01-2018-0012
8. Liao Z, Gao M, Yan F (2012) Application study of information visualization in digital library. In: Proceedings of 2012 national conference on information technology and computer science, China, pp 979–982. https://doi.org/10.2991/citcs.2012.249
9. Bertini E, Catarci T, Di Bello L, Kimani S (2005) Visualization in digital libraries. In: Hemmje M, Niederée C, Risse T (eds) From integrated publication and information systems to information and knowledge environments: essays dedicated to Erich J. Neuhold on the occasion of His 65th birthday. Springer, Berlin, pp 183–196. https://doi.org/10.1007/978-3-540-31842-2_19
10. Chen HM (2017) Real-world uses for information visualization in libraries. Libr Technol Rep 53(3):21–25. https://journals.ala.org/index.php/ltr/article/view/6291
11. Hu J, Wang D, Zhang B (2019) Research on library information visualization retrieval technology based on readers' needs. IOP Conf Ser Mater Sci Eng 569:052049. https://doi.org/10.1088/1757-899X/569/5/052049
12. Tell your library's story: create a library infographic. Library Stile. https://librarystile.com/2017/07/17/tell-your-librarys-story-create-a-library-infographic/. Accessed 23 Jul 2020
13. How scholars share journal articles: implications for the library. Library Connect. https://libraryconnect.elsevier.com/articles/how-scholars-share-journal-articles-implications-library. Accessed 23 Jul 2020
14. Beyond downloads: how scholars save & share articles. Library Connect. https://libraryconnect.elsevier.com/articles/get-infographic-beyond-downloads-how-scholars-save-share-articles. Accessed 23 Jul 2020
15. 5 Library infographics for national library card sign-up month. Easelly. https://www.easel.ly/blog/library-infographic-ideas/. Accessed 23 Jul 2020
16. US Public Libraries. http://www.ala.org/tools/sites/ala.org.tools/files/content/initiatives/plftas/issuesbriefs/issuebrief-weatherstorm.pdf. Accessed 23 Jul 2020
17. What's your library card really worth? Nashville Library. https://library.nashville.org/whats-your-library-card-really-worth. Accessed 21 May 2020
18. Gkioulekas P, Polydoratou P (2019) Information visualisation and library data: a case study of Public Library of Veria, Greece. In: ELPUB 2019 23d international conference on electronic publishing. OpenEdition Press, Marseille. https://doi.org/10.4000/proceedings.elpub.2019.7

Additional Resources

1. http://sealthreinhold.com/school/tuftes-rules/rule_one.php
2. https://guides.uflib.ufl.edu/covid-19/visualization
3. http://www.fernandaviegas.com/
4. https://www.data-to-viz.com/
5. https://www.visualisingdata.com/resources/
6. http://book.visualisingdata.com/chapter/0
7. https://textvis.lnu.se/
8. https://trustmlvis.lnu.se/
9. https://www.c82.net/work/?id=347
10. http://www.stefanieposavec.com/dear-data
11. http://notabilia.net/

Chapter 10
Tools and Techniques for Text Mining and Visualization

Abstract This chapter covers 19 popular open-access text mining and visualization tools, including R, Topic-Modeling-Tool, RapidMiner, WEKA, Orange, Voyant Tools, Gephi, Tableau Public, Infogram, and Microsoft Power BI, among others, with their applications, pros, and cons. As there are many text mining and visualization tools available, we covered only those open-source tools that have a simple GUI so that information professionals who are new to these tools can learn to use and implement them in their daily work.

10.1 Introduction

A large number of machine learning (ML) and data mining (DM) tools have been created in the past 25 years, where the primary intent was to aid the cumbersome data analysis process. These tools are devised for several purposes: analysis, prediction, recommendation systems, and processors (for images, sounds, or languages). Yang et al. [1] loosely defined "*Library 2.0* as a model for a modernized form of library service, reflecting a transition within the library world where services are delivered to users." In contrast, they defined "*Librarian 2.0* to imply a change towards a revised understanding of the core competencies and qualities of librarianship" [1]. *Library 2.0* is more focused on user-centered modifications and engagement in the formation of community and content. Recently, as *Industry 4.0* produces more software and platforms to provide tools for text and data mining (TDM), both *Web 4.0 (Intelligent Web)* and *Industry 4.0 (Intelligent Industry)* have amalgamated to give rise to what is called *Library 4.0 (intelligent library)*. An intelligent library (*Library 4.0*) evaluates significant data functions to analyze information and then uses the analyzed result for decision-making or to provide a value-based service to its customers. Figure 10.1 summarizes the features of *Web 4.0*, *Industry 4.0*, and *Library 4.0*, which are derived from the works of [2] and [3].

The development of web and digital libraries has made it relatively easier to retrieve a significant amount of textual documents that can be developed as valuable data resources. Textual data in databases and online sources raises a question about how to keep a check on the data and analyze it to provide services in *Library 4.0*.

© The Author(s), under exclusive license to Springer Nature Switzerland AG 2022
M. Lamba, M. Madhusudhan, *Text Mining for Information Professionals*,
https://doi.org/10.1007/978-3-030-85085-2_10

Fig. 10.1 Features of Library 4.0, Industry 4.0, and Web 4.0

It is not practical to efficiently extract and analyze useful information manually. Software solutions in the form of automatic tools by *Industry 4.0* are required to analyze, extract, and organize a large volume of relevant information from the textual data. Text Analysis Portal for Research (TAPoR 3)[1] and Digital Research Tools (DiRT) Directory[2] are examples of some websites that curate these *Industry 4.0* text and data mining (TDM) tools, which are freely available to use. With the increasing demands to visualize the results from the substantial amount of textual data available on the Web, information visualization is receiving considerable importance in research. In this chapter, we will explicitly explore the *big data* feature of *Library 4.0* for TDM tools to build the framework of information visualization.

10.2 Text Mining Tools

As there are many TDM tools in the market, we only covered the open-source TDM tools that have a simple GUI so that anyone can learn to use and implement them at their work or use it for their research. The description of selected open-source TDM tools is enumerated in the following subsections.

10.2.1 R

R[3] is a free software that performs statistical computing and graphics. RStudio[4] is the *integrated development environment (IDE)* for R. R is an open- and free-source TDM tool written in Java and C++. It was first released in 2011 and is available for Windows, Linux, and Mac. It is one of the powerful tools for data analysis and statistical computing. It is available as RStudio Desktop for running

---

[1] http://tapor.ca/home.

[2] http://dirtdirectory.org/.

[3] https://www.r-project.org/.

[4] https://www.rstudio.com/products/rstudio/download/.

the program locally as a regular desktop application and RStudio Server for running a program using a web browser. It supports HTML, PDF, Word documents, and slide shows. There are various packages like dplyr, tidyr, readr, data.table, SparkR, ggplot2, etc., available for different kinds of TDM problems. R can be used to perform classification, clustering, deep learning, topic modeling, sentiment analysis, bibliometrics, predictive modeling, data manipulation, data exploration, machine learning, visualization, and much more.

10.2.1.1   Pros

1. Style of coding is easy
2. Open-source
3. Availability of instant access to over 7800 packages customized for various computation tasks
4. Helpful community. Numerous forums are there to help
5. Get high-performance computing experience
6. One of highly sought skills by analytics and data science companies
7. Ease of reproducibility
8. Easier to find the mistakes and correct them
9. Can pull data from APIs, SPSS files, and many other formats
10. Can do web scraping
11. Can make web pages from the analysis
12. Robust statistical tools

10.2.1.2   Cons

1. Higher learning curve
2. More time needed on front-end analysis
3. Decentralized
4. Inconsistent
5. Complexity

10.2.2   Topic-Modeling-Tool

Topic-Modeling-Tool (TMT)[5] is a *graphical user interface (GUI)* tool for Latent Dirichlet Allocation (LDA)-based topic modeling developed in 2011 and was supported by a National Leadership Grant from the Institute of Museum and Library Services to the University of Michigan, the University of California—Irvine, and

---

[5] https://code.google.com/archive/p/topic-modeling-tool/.

Yale University. It uses Java and MALLET at the backend [4]. It uses plain text format by default. Topic-Modeling-Tool is primarily used to conduct topic modeling on a corpus of documents using LDA. The Google Code wiki [5] for the TMT emphasizes that:

> topic models provide a simple way to analyze large volumes of unlabeled text. A *topic* consists of a cluster of words that frequently occur together. Using contextual clues, topic models can connect words with similar meanings and distinguish between uses of words with multiple meanings. The GUI has two main windows: (i) basic and (ii) advanced:

1. **Basic Options**

   a. *Input*: Firstly, the location of data that is to be imported is specified. The GUI of TMT supports importing from either a single file or a directory. When the input is a single file, each line of the file is taken as a document instance. In contrast, when importing from a directory, each text file is taken to be a document instance. Do make sure that the directory contains only the text files.
   b. *Output Directory*: This is where all the output files are saved by the tool. By default, this is the current directory of the JAR file.
   c. *Number of Topics*: This option sets the total number of topics, T, for the topic modeling algorithm. The user chooses T to input the document instances, where changing T allows users to vary the granularity of the produced topics. Although there is no hard-and-fast rule to set T, it is a good idea to take into account the size of the input dataset.
   d. *Building the Topic Model*: With at least T as the basic option, click on the Train Topics button to run the topic modeling algorithm. This may take a few to several minutes to finish running, depending on the size of the dataset.
   e. *Model Output*: The output produced by the system is available to the user in mainly two forms – as comma-separated value (CSV) files and as static HTML files. Both formats essentially contain similar information, only presented differently. The CSV output consists of the following three files inside the output_csv folder:

      • Topics_Words.csv – Top words correspond to each of the T topics.
      • TopicsinDocs.csv – T topics arranged in descending order of importance for each document.
      • DocsinTopics.csv – List of top documents (max. of 500) corresponding to each of the T topics.

      The HTML output consists of an index file of topics all_topics.html, T topic web pages, and D document web pages. Using the hyperlinks in these pages, one can navigate between topics and documents in a simple and fast manner.

2. **Advanced Options**: Besides the basic options provided in the first window, there are more advanced parameters that can be set by clicking the advanced button.

   a. *Remove stopwords*: It removes a list of stopwords from the text if selected.
   b. *Stopword file*: It reads one stopword per line from a file. The default for the tool is Mallet's list of standard English stopwords.
   c. *Preserve case*: It does not force all strings to lowercase if selected.
   d. *Number of iterations*: It provides the number of iterations of Gibbs sampling to run. The default is as follows:

      • For T<100 default iterations = 200
      • For T>500 default iterations = 1000
      • Else default iterations = 2*T

   If you run for more iteration, the topic coherence may improve.

3. *Number of Topic Words Printed*: The number of most probable words to print for each topic after model estimation. The default is to print top 10 words. For this study, top 5 words were chosen.
4. *Topic Proportion Threshold*: Do not print topics with proportions less than this threshold value. A good suggested value is 5%. Users can increase this threshold for shorter documents.

10.2.2.1   Pros

1. User-friendly
2. Easy to use
3. Open-source

10.2.2.2   Cons

1. Does not let you tweak or see in between steps

10.2.3   *RapidMiner*

Nguyen et al. explain that "RapidMiner[6] (formerly YALE, Yet Another Learning Environment) is an easy-to-use visual environment for data preparation, machine learning, text mining, deep learning, and predictive analytics. It was developed in 2001 at the Artificial Intelligence Unit of the Technical University of Dortmund and is a cross-platform framework developed on the open core model written in Java" [6]. They further elucidate that "it supports interactive mode (GUI), command-line interface (CLI), and Java API. It represents a new approach to design even very complicated problems by using a modular operator concept and supports about twenty-two file formats. For large-scale data analytics, RapidMiner supports unsupervised learning in Hadoop (Radoop), supervised learning in memory with scoring on the cluster (SparkRM) and scoring with native algorithms on the cluster" [6]. RapidMiner can perform the following tasks:

1. Clustering
2. Data replacement
3. Data partitioning
4. Automation and process control
5. Data access and management
6. Data preparation
7. Bayesian modeling

---

[6] https://rapidminer.com/.

8. Data exploration
9. Visual workflow designer
10. Descriptive statistics
11. Scoring
12. Weighting and selection
13. Market basket analysis
14. Data sampling
15. Similarity calculation
16. Modeling evaluation
17. Graphs and visualization

10.2.3.1   Pros

1. Robust features
2. User-friendly interface that can perform all kinds of transformations, calcula-
   tions, joins, and filters without coding
3. Maximization of data usage
4. Students or researchers can apply for a 1-year renewable RapidMiner Educational License Program to get unlimited rows to access and many more additional
   services which are otherwise paid. Under the Educational License, users also
   get "access to free online training courses via RapidMiner Academy and free
   certification exams with credentials that can be shared on LinkedIn"[7]
5. Reads and merges databases such as SQL Server, Informix, MySQL, and Oracle
6. Has a wide set of algorithms with learning schemes, models, and algorithms from
   WEKA and R scripts
7. Strong community and cross-platform framework
8. Users can download many add-on packages from the marketplace for text
   analysis, image analysis, recommender systems, etc
9. Provides support for most types of databases, which means that users can import
   information from a variety of database sources to be examined and analyzed
   within the application

10.2.3.2   Cons

1. Proprietary product that offers solutions for more extensive problems where the
   free edition allows users to work with 10,000 rows, but if one needs processing
   of more rows, it is costly
2. Sharing analysis results is not easy
3. Not suitable for making dashboards

---

[7] https://rapidminer.com/educational-program/.

10.2.4   Waikato Environment for Knowledge Analysis (WEKA)

WEKA is implemented in Java and was developed by the University of Waikato, New Zealand, in 1977. It is licensed under the General Public License (GNU) for knowledge analysis. It can be downloaded for Windows, Mac, and Linux. It has a wide set of algorithms for TDM tasks such as data pre-processing, regression, classification, association, clustering, deep learning, and visualization. The algorithms can directly be applied to a dataset or using Java. It reads and uses data in ARFF format by default. WEKA has five application interfaces:

1. *Explorer*: It has a graphical interface to perform data mining tasks on raw data and contains six tabs, namely, preprocess, classify, association, cluster, select attributes, and visualize.
2. *Experimenter*: It allows users to execute different experimental variants on datasets.
3. *Knowledge Flow*: It executes *explorer* application with drag-and-drop feature and also supports incremental learning from previous results.
4. *Simple CLI*: It is the command-line interface for executing commands from a terminal.
5. *Workbench*: It combines all the applications and converts all the GUI interfaces into one.

10.2.4.1   Attribute Relation File Format (ARFF)

ARFF constitutes two sections: (a) header and (b) data. The header section describes the relation or the attribute name, dataset name, and the type, whereas the data section lists the data instances. It needs the declaration of the attribute, data, and relation.

10.2.4.2   Pros

1. Platform independent
2. Open-source
3. Offers different machine learning algorithms for different TDM approaches
4. Easy to use
5. Provides flexibility for scripting
6. Portability
7. Best suited for mining association rules
8. Strong in machine learning techniques
9. Suitable for developing new machine learning schemes
10. Can be integrated into other Java packages

10.2.4.3 Cons

1. Not capable of multi-relational data mining
2. Does not support sequence modeling
3. Lacks proper and adequate documentation and suffers from "Kitchen Sink Syndrome" where systems are updated constantly
4. Worse connectivity to Excel spreadsheet and non-Java-based databases
5. CSV reader not as robust as in RapidMiner
6. Weak in classical statistics
7. Does not have the facility to save parameters for scaling to apply to future datasets
8. Does not have an automatic facility for parameter optimization of machine learning/statistical methods

10.2.5 Orange

Orange[8] is an open-source machine learning, data mining, and data visualization tool. It was first developed by the University of Ljubljana in 1996 and is written in Python and C++. It has a cross-platform operating system and can be downloaded for Mac, Windows, or Linux operating system. It is also available through Anaconda Navigator. It can read files in native or any other data format. It contains tools for experiment selection, recommendation systems, and predictive modeling. It can be used in different disciplines, for instance, biomedicine, business studies, bioinformatics, genomic research, geo-informatics, digital humanities, social sciences, etc. It has interactive data visualization that uncovers hidden data patterns and provides intuition behind the data analysis procedure. Its components are called widgets to create a workflow by linking predefined or user-designed widgets on the Canvas (graphical interface). It has six widgets sets, viz., data, classify, evaluate, regression, unsupervised, and visualize. Data fusion, text mining, and bioinformatics widgets are available as add-ons. Different sub-widgets of the primary widgets are as follows:

1. *Data*: It contains widgets for data input, data filtering, sampling, imputation, feature manipulation, and feature selection.
2. *Visualize*: It comprises widgets for common visualization such as box plot, histograms, scatter plot, etc.; model-specific visualizations like dendrogram, silhouette plot, tree visualization, etc.; and multivariate visualization such as mosaic display, sieve diagram, etc.
3. *Classify*: It has a set of supervised machine learning algorithms for classification.

---

[8] https://orangedatamining.com/.

4. *Regression*: It consists of a set of supervised machine learning algorithms for regression.
5. *Evaluate*: It constitutes cross-validation, sampling-based procedures, reliability estimation, and scoring of prediction methods.
6. *Unsupervised*: It is composed of unsupervised learning algorithms for clustering (k-means, hierarchical clustering) and data projection techniques (multidimensional scaling, principal component analysis, correspondence analysis).
7. *Associate*: It has widgets for mining frequent itemsets and association rule learning.
8. *Bioinformatics*: It contains widgets for gene set analysis, enrichment, and access to pathway libraries.
9. *Data fusion*: It consists of widgets for fusing different datasets, collective matrix factorization, and exploration of latent factors.
10. *Educational primary widget*: It has widgets for teaching machine learning concepts, such as k-means clustering, polynomial regression, stochastic gradient descent, etc.
11. *Geo*: It is composed of widgets for working with geospatial data.
12. *Image analytics*: It works with images and ImageNet embedding widgets.
13. *Network*: It comprises widgets for graph and network analysis.
14. *Text Mining*: It constitutes widgets for natural language processing and text mining.
15. *Time series*: It has widgets for time-series analysis and modeling.
16. *Spectroscopy*: It consists of widgets for analyzing and visualization of (hyper)spectral datasets.

10.2.5.1   Pros

1. Open-source
2. Featured on the front page of Anaconda Navigator
3. Interactive data visualization
4. Can perform both simple and complex data analyses
5. Visual programming
6. Supports hands-on training and visual illustrations
7. Availability of various add-ons for extended functionality
8. Easiest tool to learn
9. Has better debugger
10. Scripting data mining categorization problems is more straightforward in Orange

10.2.5.2   Cons

1. Does not give optimum performance for association rules
2. Limited list of machine learning algorithms

3. Machine learning is not handled uniformly between the different libraries
4. Weak in classical statistics; although it can compute basic statistical properties of the data, it provides no widgets for statistical testing
5. Reporting capabilities are limited to exporting visual representations of data models

10.2.6   Voyant Tools

Voyant Tools[9] is a web-based text reading and analysis environment under CC 4.0. It was first released in 2003. It has a cross-platform operating system and is available in 10 languages. It can work with different formats such as HTML, plain text, PDF, XML, MS Word, and RTF. It can perform text analysis, statistical analysis, and data mining. It has five preliminary tools: *Cirrus*, *Reader*, *Trends*, *Summary*, and *Contexts*, and additional tools, viz. *Terms*, *Links*, *Collocates*, *Documents*, *Phrases*, and *Bubblelines*, are also accessible by clicking the tabs in each tool pane. The function of each tool at Voyant Tools has been summarized below:

1. *Bubblelines*: It visualizes the frequency and distribution of terms in a corpus.
2. *Bubbles*: It is a playful visualization of term frequencies by the document.
3. *Cirrus*: It is a word cloud that visualizes the top frequency words of a corpus or document.
4. *Collocates Graph*: It represents keywords and terms that occur nearby as a force-directed network graph.
5. *Corpus Collocates*: It is a table view of which terms appear more frequently in proximity to keywords across the entire corpus.
6. *Contexts (or Keywords in Context)*: It shows each occurrence of a keyword with a bit of surrounding text (the context).
7. *Correlations*: It enables an exploration of the extent to which term frequencies vary in sync (terms whose frequencies rise and fall together or inversely).
8. *Document Terms*: It is a table view of document term frequencies.
9. *Corpus Terms*: It is a table view of term frequencies in the entire corpus.
10. *Documents*: It shows a table of the documents in the corpus and includes functionality for modifying the corpus.
11. *Knots*: It is a creative visualization that represents terms in a single document as a series of twisted lines.
12. *Mandala*: It is a conceptual visualization that shows the relationships between terms and documents.
13. *MicroSearch*: It visualizes the frequency and distribution of terms in a corpus.
14. *Phrases*: It shows repeating sequences of words organized by frequency of repetition or number of words in each repeated phrase.

---

[9] https://voyant-tools.org/.

15. *Reader*: It provides a way of reading documents in the corpus. The text is fetched on demand as needed.
16. *StreamGraph*: It is a visualization that depicts the change of the frequency of words in a corpus (or within a single document).
17. *Summary*: It provides a simple, textual overview of the current corpus, including information about words and documents.
18. *TermsRadio*: It is a visualization that depicts the change of the frequency of words in a corpus (or within a single document).
19. *TextualArc*: It is a visualization of the terms in a document that includes a weighted centroid of terms and an arc that follows the terms in document order.
20. *Topics*: It provides a rudimentary way of generating term clusters from a document or clusters from a document or corpus and then seeing how each topic (term cluster) is distributed across the document or corpus.
21. *Trends*: It shows a line graph depicting the distribution of a word's occurrence across a corpus or document.
22. *Veliza* is an experimental tool for having a (limited) natural language exchange (in English) based on your corpus.
23. *Word Tree*: It is a tool that allows you to explore how words are used in phrases.

10.2.6.1 Pros

1. Valuable and powerful tool that is employable by both beginner and advanced user
2. Open-source
3. User-friendly
4. Simple user interface
5. Compatible with a wide range of file formats
6. Offers several tools for analysis
7. Can create a persistent link that can be used to retrieve the complete Voyant Tools dashboard and the corpus
8. Individual modules can be exported, and, in some modules, data can be exported as an image, tabular data in plain text, CSV, or tab-delimited formats

10.2.6.2 Cons

1. Still available in beta mode
2. Platform can be slow to load or gets hung
3. Scrolling through tools is not always a smooth and straightforward process
4. Favorite feature list cannot be saved for future use when generating URL
5. Challenge of gathering information using some visualization tools (e.g., Knots, Lava, and Mandala)
6. Cannot create an account; thus, cannot retrieve the older visualizations

### 10.2.7	Science of Science (Sci2) Tool

Sci2 Tool[10] is a mapping and visualization tool "specifically designed for the study of science. It is built on the Cyberinfrastructure Shell (CIShell). It is an open-source software framework for the easy integration utilization of datasets, algorithms, tools, and computing resources developed by Cyberinfrastructure for Network Science Center (http://cns.iu.edu) at Indiana University" [7]. It supports 14 file formats and could be used to visualize and analyze networks for co-authors, co-citations, document, and journal bibliographic coupling. It supports four kinds of analyses, viz., temporal, geospatial, topical, and network, at three levels, viz., micro (individual), meso (local), and macro (global). "Users of the tool can (i) access science datasets online or load their own, (ii) perform different types of analysis with the most effective algorithms available, (iii) use different visualizations to explore and understand specific datasets interactively; and (iv) share datasets and algorithms across scientific boundaries" [7]. It has six major features:

1. *Loading data* loads supported file format for preparation, analysis, or visualization.
2. *Data preparation* extracts network from raw data or updates currently existing networks by merging nodes and removing duplicates.
3. *Processing* cleans data for analysis and visualization.
4. *Analysis* employs a variety of advanced algorithms for temporal, topical, geospatial, and network data.
5. *Modeling* generates a graph with aging, scaling, random, and other specifications.
6. *Visualization* visualizes temporal, topical, geospatial, and network data.

#### 10.2.7.1	Pros

1. Explicitly designed for Science of Science (SoS) research and practice, well documented, and easy to use
2. Empowers many to run common studies while making it easy for experts to perform novel research
3. Advanced algorithms, effective visualizations, and many (standard) workflows
4. Supports micro-level documentation and replication of studies
5. Open-source
6. Supports network overlay for geographical map

---

[10] http://cns.iu.indiana.edu.

10.2.7.2   Cons

1. Offers only the US or world map and not country-specific maps for network analysis
2. GUESS visualization is slow
3. Typing commands into the interpreter can be tricky

10.2.8   LancsBox

LancsBox[11] is a Lancaster University corpus toolbox that is primarily used to analyze data in various languages and currently supports 20 languages. It runs on Java and can be downloaded for Windows, Mac, or Linux operating system. The LancsBox User Manual [8] states that:

> It supports many different formats and uses multiple third-party tools and libraries such as Apache Tika, Gluegen, Groovy, JOGL, minlog, QuestDB, RSyntaxTextArea, smallseg, and TreeTagger. Significant features of LancsBox include:
>
> 1. It uses powerful searching experiences at different levels of corpus annotation using *simple*, *wildcard*, *smart*, and *regex* searches;
> 2. *KWIC tool* generates a list of all instances of a search term in a corpus in the form of a concordance which can then be used to (i) find the frequency of a word or phrase in a corpus, (ii) find frequencies of different word classes such as nouns, verbs, adjectives, (iii) find complex linguistic structures such as passives, split infinitives, etc., using *smart* searches, and (iv) sort, filter, and randomize concordance lines;
> 3. *Whelk tool* provides information about how the search term is distributed across corpus files and can be used to (i) find absolute and relative frequencies of the search term in corpus files, (ii) filter the results according to different criteria, and (iii) sort files according to absolute and relative frequencies of the search term;
> 4. *Word tool* allows in-depth analysis of frequencies of types, lemmas, and Part of Speech (POS) categories as well as comparison of corpora using the keywords technique and can be used to (i) compute frequency and dispersion measures for types, lemmas, POS tags, (ii) visualize frequency dispersion in corpora, (iii) compare corpora using the keyword technique, and (iv) visualize keywords;
> 5. *GraphColl tool* identifies collocations and displays them in a table or as a collocation graph or network. It can be used to (i) find collocates of a word or phrase, (ii) find colligations, i.e., co-occurrence of grammatical categories, (iii) visualize collocations and colligations, (iv) identify shared collocates of words or phrases, and (v) summarise discourse in terms of its 'aboutness';
> 6. *Text tool* enables in-depth insight into the context for the used words or phrases and can be used to (i) view a search term in full context, (ii) preview a text, (iii) preview a corpus as a run-on text, (iii) check different levels of annotation of a text;
> 7. *Ngram tool* allows in-depth analysis of frequencies of n-gram types, lemmas, and POS categories as well as comparison of corpora using the key n-gram technique and can be used to (i) compute frequency and dispersion measures for n-gram types, lemmas, and

---

[11] http://corpora.lancs.ac.uk/lancsbox/download.php.

POS tags, (ii) visualize frequency and dispersion in corpora, (iii) compare corpora using the key n-gram technique, and (iv) visualize key n-gram;

8. *Wizard tool* combines the power of all tools in the LancsBox. It searches corpora and produces research reports. It can be used to (i) carry out simple or complex research, (ii) produce a draft report, and (iii) download all relevant data.

10.2.8.1 Pros

1. Can be used for various automated text pre-processing processes such as POS tagging and n-grams
2. Can be used by historians, sociologists, educators, linguists, or anyone interested in the language
3. Searches corpora quickly
4. Can be used to compare multiple corpora in different languages
5. Visualizes the data
6. Analyzes data in any language
7. Automatically annotates data for parts-of-speech tagging

10.2.8.2 Cons

1. Gloomy and clunky interface
2. Does not offer proper configuration options

10.2.9 ConText

Connections and **Text**s (ConText)[12] facilitates (i) extraction of network data from text data, that is, *relation extraction*, and (ii) joint analysis of text and network data. It is built in Java and can run on PCs and Macs. It accepts only plain text (.txt) files as input where results are not overwritten and modified data becomes new datasets. It has varieties of open-source libraries. Its main features include:

1. *Summarization*—gains quick understanding and overview of the data using topic modeling, sentiment analysis, and corpus statistics. It uses Mallet[13] for topic modeling, which is based on LDA. For sentiment analysis, ConText uses a predefined sentiment lexicon, based on POS, and maps text terms to positive, negative, and neutral categories. Users can modify the predefined lexicon or use their own lexicon, including their own choice of categories. In corpus statistics,

---

[12] http://context.ischool.illinois.edu/.
[13] http://mallet.cs.umass.edu/topics.php.

it gives an overview of (weighted) word frequencies and distribution of words across texts.

2. *Text Pre-processing*—prepares the data for further analysis using POS, stopword removal, stemming, and bigram detection.
3. *Network Construction*—extracts network data from text data from entity types, co-occurrence, and syntax.

10.2.9.1   Pros

1. Easy to use
2. User-friendly GUI
3. POS can be performed in English, Arabic, Chinese, French, and German language text data
4. Integrates both text analysis and network analysis

10.2.9.2   Cons

1. Cannot use files other than those in plain text format

10.2.10   Overview Docs

Overview Docs[14] is a web-based application that helps to read and analyze thousands of documents quickly. It can import many formats and languages and requires an account to use the tool. Only the users can view documents that they uploaded, but still, they should be careful with any private data. It offers customization of the tool through API. The significant features of the tool include built-in OCR (optical character recognition), full-text search, visualization, entity detection, clustering, tagging, and metadata services. The Overview Docs public server was decommissioned on Aug. 1, 2021. You can run Overview Docs in your computers on your local browser by following instructions from https://github.com/overview/overview-local.

10.2.10.1   Pros

1. Easy to use
2. Visual workflow
3. Can export tagged files as CSV, Excel, or Zip

---

[14] https://www.overviewdocs.com/.

4. Beneficial for the organization of multiple files and folders using clustering, tagging, and entity detection techniques
5. Suitable for initial exploratory analyses of the corpus

10.2.10.2   Cons

1. Cannot export word cloud and clustering (tree) visualizations
2. Does not offer any advanced techniques for text analysis or visualization

10.3   Visualization Tools

We will now focus on various open-source visualization tools that can be used to visualize the results after conducting a TDM process. Some of the important data visualization tools are as follows.

10.3.1   Gephi

"Gephi[15] is an interactive visualization and exploration platform for all kinds of networks and complex systems, dynamic and hierarchical graphs" [9]. It was first released in 2008 and is written in Java and OpenGL. It is an open-source and free software and can run on Windows, Linus, and MacOS operating systems. It supports the opening of network files from 11 formats. It can visualize exploratory data analysis, link analysis, social network analysis, biological network analysis, and poster creation. It uses betweenness centrality, closeness, diameter, clustering coefficient, PageRank, community detection (modularity), random generators, and shortest path statistics and metrics framework. Users can also visualize a network over time using Gephi.

10.3.1.1   Pros

1. No programming skills needed
2. High performance
3. Have customizable plugins
4. Easy to use
5. Easy to edit, drag, and customize
6. Flexible in changing the color and size of the graph

---

[15] https://gephi.org/users/download/.

7. Basic support for graph clustering
8. Supports dynamic filtering
9. Open software

10.3.1.2   Cons

1. Graph navigation is not good
2. Few visual glitches
3. No linking among views

10.3.2   *Tableau Public*

Tableau Public[16] is an interactive data visualization tool that connects and extracts data from many sources. The data pulled from a file can either be connected live or be extracted to the Tableau Desktop. Users need to create an account in Tableau Public to start working with it. It has both desktop and online versions where one does not need to require any technical or programming skills to visualize the data. Some of the significant features of Tableau include data blending, real-time analysis, and collaboration of data.

10.3.2.1   Pros

1. 10GB storage for free
2. Quick startup
3. Ease of use
4. High performance
5. Multiple data source connections
6. Mobile-friendly
7. Remarkable visualization capabilities
8. Security of data as you can either allow public view or not

10.3.2.2   Cons

1. Does not allow data to be kept private
2. Embedment issues
3. Data source limitations as users will not be able to connect to their SQL database and are limited to Excel, Text, and Access files and data

---

[16] https://public.tableau.com/en-us/s/download

4. Not able to automatically refresh the data
5. Poor versioning

10.3.3   Infogram

Infogram[17] helps to generate infographics for user's data. It is a web-based application. By combining elements like text image, chart, diagram, and video, Infogram presents an effective tool for data presentation. It helps to explain complex issues easily and gives better insight into the data and presents complex data in a concise and highly visual way. Users can use it to tell data stories effectively by making information easy to digest, educational, and engaging.

10.3.3.1   Pros

1. Ability to connect graphs and charts to live data sources like Google Sheets or Dropbox
2. Can choose from over 35 charts and 550 maps
3. Can visualize with complex data
4. Ability to design interactive visualization through toggled datasets, embedded videos, and SlideShare integration
5. Multiple professional templates
6. Offers embedded codes for web pages
7. Can export visualization in several formats such as PNG, JPEG, GIF, PDF, and HTML

10.3.3.2   Cons

1. Does not include rich text editing, i.e., users cannot edit font size or style
2. Fewer built-in data sources compared to other apps

10.3.4   Microsoft Power BI

The Data-Flair blog specifies that Microsoft Power BI[18] "is a collection of business intelligence tools such as software services, apps, and data connectors. It is a cloud-based platform that consolidates data from varied sources into a single dataset. It

---

[17] https://infogram.com/.
[18] https://powerbi.microsoft.com/en-us/downloads/.

can be used for data visualization, evaluation, and analysis by making shareable reports, dashboards, and apps" [10]. The blog further describes that "Microsoft offers three types of Power BI platforms: (i) desktop application, (ii) SaaS (Software as a Service), and (iii) mobile. It can import data from local databases, cloud-based data sources, big data sources, Excel files, and other hybrid sources" [10]. Significant features of Power BI include:

1. *Attractive Visualizations*: It offers a wide range of detailed and attractive visualizations. It also has a library for custom visualizations.
2. *Get Data (Data Source)*: It helps users to select data sources anywhere from on-premise to cloud-based and from unstructured to structured. It adds new data sources every month.
3. *Dataset Filtration*: Users can filter datasets into smaller subsets.
4. *Customizable Dashboards*: It is composed of multiple visualizations as tiles that are generated from single pages of reports. These dashboards are shareable as well as printable.
5. *Flexible Tiles*: A tile is a single block containing a visualization in the Power BI dashboard. These tiles can be placed anywhere on the dashboard, and their size can be changed.
6. *Navigation Pane*: Navigation pane has options of datasets, dashboards, and reports where users can navigate between all three as per their convenience.
7. *Informative Reports*: In Power BI, reports combine dashboards with a different kind of visualization. It shows a complete and structured presentation of data in different ways and reveals essential insights from the data.
8. *Natural Language Q&A Question Box*: It is a unique feature of Power BI where users can ask questions in natural language to search for available data and information.
9. *DAX Data Analysis Function*: There are about 200 data analysis expressions which are predefined codes to perform specific functionalities on data.

10.3.4.1 Pros

1. Powerful tool for data analysis and visualization
2. Can process approximately 2GB of data at a time
3. User-friendly tool with impressive drag-and-drop features
4. Easily shareable reports
5. Power BI Desktop version is free of cost
6. Customizable visualizations
7. Excel integration
8. Data connectivity
9. Reports can be embedded in web-based or other apps
10. All the imported data is stored in a centralized location which means users can access the data anytime, from anywhere, and from any platform as many times as they want
11. Interactive visualizations

10.3.4.2   Cons

1. To use additional Power BI services and publish reports on the cloud, users have to subscribe to the paid version of the Power BI cloud service
2. Cannot handle complex relationships between tables
3. Have a crowded interface
4. Rigid formulas
5. To process large data volumes beyond 2GB, users have to subscribe to the paid version
6. Complex to understand and master

10.3.5   Datawrapper

Datawrapper[19] is an innovative web-based data visualization tool developed by ABZV. It enables users to create simple yet interactive charts and graphs that can be embedded online. It supports PDFs, Excel, Google Sheets, and CSV formats in addition to copied-and-pasted data from the applications. It offers 19 chart types, 3 map types, and data tables.

10.3.5.1   Pros

1. No code or design skills required
2. Interactive, responsive, and embeddable visualizations
3. The process of creating visualization is rapid and simple
4. Can customize and annotate the charts
5. Visualizations can be exported in various formats
6. The free version supports 10,000 monthly chart views
7. Hosts a blog called *CHARTABLE*,[20] where they regularly write about best practices of data visualization
8. Includes a built-in color-blindness checker

10.3.5.2   Cons

1. Data is stored on Datawrapper server with a free plan
2. Storing data on the user's server can be complicated
3. Customization of fonts and colors can be difficult
4. Allows basic interactive visualizations only

---

[19] https://app.datawrapper.de/chart/s7Vf7/upload.

[20] https://blog.datawrapper.de/.

5. Does not have advanced data cleaning options
6. Limited data sources

10.3.6 RAWGraphs

RAWGraphs[21] is a web-based data visualization tool developed by DensityDesign Research Lab and Calibro. It is used to create "vector-based visualizations on top of the d3.js library. It only works with tabular data (such as spreadsheets and CSV) as well as with copied-and-pasted texts from other applications. The visualizations can be embedded online" [11].

10.3.6.1 Pros

1. Visualization can be easily edited for further refinements
2. Data is not stored on the RAWGraphs server
3. Highly customizable and extensible
4. Options to make customizable visualizations are available
5. Can run locally on the user's machine

10.3.6.2 Cons

1. Knowledge of programming is required to make customized visualizations
2. Most of the charts are for advanced data visualization purposes

10.3.7 WORDij

WORDij[22] is a free semantic network tool for capturing relationships between words and their assigned word-pair link strengths. It is written in C programming language, whereas Java is used to build its GUI. It can run on Mac and Windows operating systems and can analyze files in UTF-8 format. The output files from WORDij can be imported to eight other network analysis programs, including UCINET, NodeXL, Pajek, and Negopy. The tool computes the similarity of pairs of networks from different sets of text using quadratic assignment procedures (QAP) that produce a correlation coefficient for the comparison of whole networks. Its features include:

---

[21] https://app.rawgraphs.io/.
[22] https://www.wordij.net/download.html.

1. *WordLink* counts words and word pairs, and the results can be used by other modules.
2. *QAPNet* calculates an overall measure of the similarity of two whole networks using a correlation coefficient from $+1$ to -1.
3. *Z-Utilities* compares two text files and determines the significant differences between the words or the word pairs.
4. *VISij* is a graphic visualization module. If multiple files are included, an animation will play a network sequence change from one file to another.
5. *OptiComm* produces messages to move two words closer, to move them further apart, or to reinforce aspects of the semantic networks.
6. *Utilities* is a proper noun extraction and a TimeSegs program for overtime analysis using input from Lexis/Nexis or NewsBank.
7. *Conversions* converts WORDij files for use with MultiNet/Negopy UCINET, NetDraw, and Pajek.

10.3.7.1 Pros

1. Can run data files as large as 550MB
2. Users can convert multi-word phrases to a single unigram such as New York City can be rendered as new_york_city
3. Free for noncommercial academic research. Commercial licensing is also available
4. Output from WORDij can be used to compute word embeddings to perform Word2Vec

10.3.7.2 Cons

1. Cannot customize the colors of the graph

10.3.8 Palladio

Palladio[23] is a web-based application for visualizing complex and multidimensional data in the form of maps and networks. It was built by the Digital Humanities Lab at Stanford University and was funded by the National Endowment for the Humanities (NEH). It only allows data to be imported in delimited-separated values format, including commas, semicolons, and tabs, or can be copied and pasted directly into the tool. Users can combine maps and timelines, filter the data, make bimodal

---

[23] http://hdlab.stanford.edu/palladio/.

network graphs, create relationships among tables, and download the data model using Palladio. The major features of the tool include:

1. *Map View* converts the coordinate data as points on the map.
2. *Graph View* helps to visualize the relationships between any two dimensions of the data.
3. *List View* arranges the dimensions of the data in a customized list.
4. *Gallery View* displays data in a grid setting for quick reference.

10.3.8.1 Pros

1. Allows to ask and answer varieties of research questions
2. Robust
3. Allows advanced data visualization

10.3.8.2 Cons

1. Cannot embed visualizations in websites
2. Cannot export the visualizations
3. Lack of deep customization
4. Cannot create an account; thus, it cannot retrieve older visualizations
5. Some of the networking visualizations are overwhelming and difficult to read
6. Does not support basic data visualization
7. Designed for exploration of data

10.3.9 Chart Studio

Chart Studio[24] is a powerful Plotly's web-based online visualization tool. It is built on D3.js and allows users to create WebGL. The data can be copied and pasted or imported from Excel or CSV files.

10.3.9.1 Pros

1. The basic version is free to use, whereas the enterprise cloud versions are paid
2. The basic cloud version is also free but requires the organization's email id and allows many features such as 100 API calls/day for Python and R, sharing of

---

[24] https://plotly.com/chart-studio/.

both data and charts up to 1000 views/day, unlimited chart creation, visualization exportation in PNG and JPEG, and import of 500KB of data

3. Can export the visualizations
4. Can embed static or fully interactive visualizations
5. Can custom templates and themes

10.3.9.2   Cons

1. Customer service is not very responsive

References

1. Yang Q, Zhang X, Du X, Bielefield A, Liu Y (2016) Current market demand for core competencies of librarianship—a text mining study of American Library Association's Advertisements from 2009 through 2014. Appl Sci 6(2):48. https://doi.org/10.3390/app6020048
2. Lee J, Lapira E, Bagheri B, Kao H (2013) Recent advances and trends in predictive manufacturing systems in big data environment. Manuf Lett 1(1):38–41. https://doi.org/10.1016/j.mfglet.2013.09.005
3. Noh Y (2015) Imagining Library 4.0: creating a model for future libraries. J Acad Librariansh 41(6):786–797. https://doi.org/10.1016/j.acalib.2015.08.020
4. Abinaya G, Winster SG (2014) Event identification in social media through latent dirichlet allocation and named entity recognition. In: Proceedings of IEEE international conference on computer communication and systems ICCCS14, pp 142–146. https://doi.org/10.1109/ICCCS.2014.7068182
5. Google Code Archive. Long-term storage for Google Code Project Hosting. https://code.google.com/archive/p/topic-modeling-tool/wikis/TopicModelingTool.wiki. Accessed 12 Aug 2020
6. Nguyen G, Dlugolinsky S, Bobák M, Tran V, López García Á, Heredia I, Malík P, Hluchý L (2019) Machine learning and deep learning frameworks and libraries for large-scale data mining: a survey. Artif Intell Rev 52:77–124. https://doi.org/10.1007/s10462-018-09679-z
7. Sci2 Team (2009) Science of Science (Sci2) Tool. Indiana University and SciTech Strategies. https://sci2.cns.iu.edu
8. LancsBox Manual (2020) http://corpora.lancs.ac.uk/lancsbox/docs/pdf/LancsBox_5.1_manual.pdf. Accessed 2 Apr 2021
9. Gephi (2017) https://gephi.org/users/download/. Accessed 12 Aug 2020
10. Power BI Tutorial (2021) https://data-flair.training/blogs/power-bi-tutorial/. Accessed 2 Apr 2021
11. RAWGraphs (2021) https://rawgraphs.io/. Accessed 2 Apr 2021

Chapter 11
Text Data and Mining Ethics

Abstract Before leaping to the critical legal and ethical issues related to text mining, it is vital to comprehend (i) the importance of data management for text mining, (ii) the lifecycle of research data, (iii) data management plan that strategizes the various data security, legal, and ethical constraints, (iv) data citation, and (v) data sharing. This chapter covers all the above-stated concepts in addition to legal and ethical issues related to text mining (such as copyright, licenses, fair use, creative commons, digital management rights), algorithm confounding, and social media research. It further presents text mining licensing conditions by selected prominent publishers and a "do's and dont's" list to help library professionals conduct text mining efficiently.

11.1 Text Data Management

In order to conduct research responsibly, it is vital to comprehend the role of data management to reuse, share, and store textual data. Data management is the everyday handling task of research data starting from the active phase of a project to its completion. It includes activities such as planning, documenting, and formatting of data and practices that support access, long-term preservation, and data usage. When a research project has been completed, the data could be used to answer additional questions not considered during the original project and to extend the study over time to perform longitudinal comparisons. It becomes very difficult if you have not effectively managed your data. Also, many publishers, research institutions, and funding agencies ask how the researchers will perform the task of data management during and after the completion of a research project by demanding more transparency in research projects and data management processes.

11.1.1   Plan

Data management plan (DMP) is a formal document that explains how the data will be generated, collected, used, managed, described, preserved, and stored for future access for a research project. It summarizes the data management strategies that will be implemented both during and after completing a research project. A DMP should include information on:

- Data description (metadata) and formats
- Handling of data (during and after the project ends)
- Data collection, processing, and generation
- If the data will be shared in open-access
- Data security and other legal or ethical constraints
- How the data will be curated and preserved

The funding agencies require grantees to submit DMP because they provide assurances that the data will be available for use for a very long time. This is important for two primary reasons:

1. It supports transparency and openness.
2. It provides greater returns on public investments in research by making the data discoverable, accessible, and reusable.

There are two open-access tools, DMPTool[1] and DMPonline,[2] among others for DMP, which researchers and information professionals can use to create DMPs. These tools provide guidance and templates for creating DMP in compliance with institutional and different international funder requirements.

11.1.2   Lifecycle

A research data lifecycle represents various activities related to data occurring throughout a research process. When performing any study which deals with data, it is essential to think about various aspects related to the management of data processed and produced in the process. Research lifecycle models help from planning to archiving of the project for proper data management. Each stage produces a specific data product in the data lifecycle that requires a range of considerations, responsibilities, and activities. There are numerous data lifecycle models from simple to complex, each with a different focus or perspective. Some of the selected lifecycle models are illustrated in Figs. 11.1, 11.2 and 11.3. These plans maybe a formal document required for submission for a grant or woven into

---

[1] https://dmptool.org/.

[2] https://dmponline.dcc.ac.uk/.

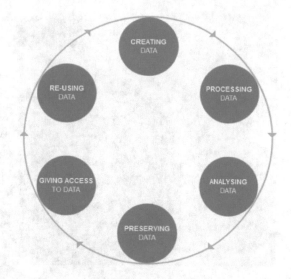

lab protocol, or maybe more informal. Regardless of whether a DMP is required or not for a research project, taking time to put it together will be beneficial down the line and save a lot of time and frustration later. In a typical research data lifecycle (Fig. 11.4), data undergo the subsequent stages of processing, analyzing, preserving, accessing, and reusing, which are identified as:

- *Stage 1: Discovery and Planning* phase helps to identify the data type and format even before commencing the data collection process. This stage may involve collecting new data, combining existing data with the new data, or analyzing the existing data. Any sensitive data such as name, phone number, location, etc., should be removed and handled ethically. Further, metadata standards, the type of documentation generated, and identification of potential users of the data should be considered and finalized in this phase. Finally, it is essential to determine the potential cost of data management, which involves formatting, documenting, storing, cleaning and anonymizing, and archiving.

  For information professionals, this stage offers an opportunity to help the researchers to develop a thorough data management plan for their research project. Librarians can assist researchers with the integration of datasets from different repositories and provide hands-on training on locating data resources and tools for managing, manipulating, and viewing datasets. They can assist in data wrangling, interpreting codebooks and other documentation, and troubleshooting problems with a dataset. This is where information professionals can be of service, leveraging their knowledge of organizational systems and preservation methods by asking questions and offering suggestions, which help to ensure that the research project begins and moves forward smoothly.

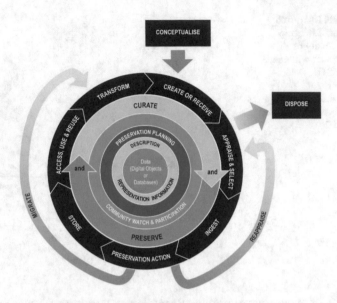

Fig. 11.2 Digital Curation Centre (DCC) Lifecycle Model (UK) (©2020 Digital Curation Centre—reprinted with the permission of Digital Curation Centre. https://www.dcc.ac.uk/guidance/curation-lifecycle-model. Accessed 18 Aug 2020)

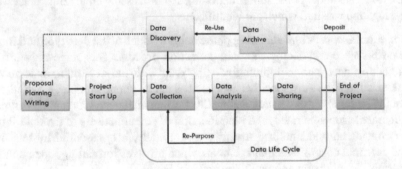

Fig. 11.3 University of Virginia Library Research Data Lifecycle (©2020 The Data Management Consulting Group, University of Virginia Library—reprinted with the permission of University of Virginia Library. https://data.library.virginia.edu/data-management/, August 17, 2020)

- *Stage 2: Initial Data Collection* phase ensures the implementation of the best data management practices which include backup and storage strategies, file organization, deciding on file organization schemes, and quality assurance protocols, including naming conventions and file versioning policies. Managing the data in the above way will make this phase effortless and decrease duplicates

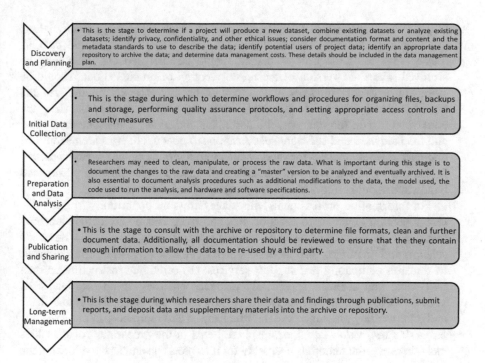

Fig. 11.4 Data management across the research lifecycle

and out-of-sync versions of the data. Proper backup and storage approaches will defend against data loss. The draft secures backup and storage protocols and quality assurance protocols. Further, one should consider data security and access controls during this phase and take appropriate steps to ensure that data is safely stored and accessible only to authorized individuals.

For information professionals, this stage offers an opportunity to advise the researchers on how the data should be organized. These include conventions for file naming, persistent unique identifiers (i.e., digital object identifiers or DOI), and versioning control. Information professionals can work with researchers to provide strategies for preventing errors from entering data and individual consultations or group training sessions on activities for both monitoring and maintenance of the quality of the data during the research project. Further, they can help the researchers to make sure that (a) the data is protected against accidental data loss by backing them up regularly; (b) appropriate data access controls are in place, mainly if researchers are dealing with sensitive data; (c) only authorized research personnel have access to the data; (d) the schedule and the responsibility for data backup are included in DMP; and (e) the data can be found and understood by others by helping the researchers with data description activities.

- *Stage 3: Preparation and Data Analysis* phase helps in cleaning, manipulating, processing, performing statistical analysis, and visualization of the data. A master version of the processed data should be analyzed and eventually be archived. The final version of the master version of processed data should be stored in read-only mode so that it cannot be inadvertently altered. All the modifications performed to the data, the model used to analyze the data, the code used to run the analysis, the acronyms and units of measurement used, and the hardware and software specifications to conduct the research should be documented meticulously for validity and reproducibility of the research process. In the case of a big data project, enough data storage, computing power, and bandwidth should be considered for processing and analysis.

  For information professionals, this stage offers an opportunity to guide the researchers on how best to describe data and files. This could include teaching them how to write readme style metadata, identifying an existing metadata schema, and creating a data dictionary. They can assist by (a) offering referrals to research computing and support services; (b) providing instruction on data processing and analysis, specifically guidance on using computing language, such as Rand Python, and statistical environments like Stata, SPSS, and SAS; and (c) providing assistance in data visualization and representation.

- *Stage 4: Publication and Sharing* phase helps in the preparation of data files and other research materials necessary for interpretation and usage of data in the future.

- *Stage 5: Long-Term Management* phase dictates that data should be stored and made available for sharing in a trustworthy data repository open to the public. Mendeley Data,[3] Registry of Research Data Repositories,[4] DataCite,[5] and FAIRsharing[6] are examples of such data repositories. The selection of an appropriate data archival repository may also depend on data type and research discipline.

  One should consult with information professionals or data repository staff for guidance while preparing the data for sharing. The specific needs and requirements of the repository, the ability to reuse the data, and proper standard metadata should be applied before submitting the data in a repository. Information professionals can assist the researchers in providing data citation and persistent DOIs for their datasets.

All research projects do not necessarily have to go through all the stages of the data lifecycle or go through all the stages in a particular order. A data lifecycle is a valuable tool for understanding the typical path of research data. Researchers may contact information professionals for their help in (a) locating potentially useful

---

[3] https://data.mendeley.com/.

[4] https://www.re3data.org/.

[5] https://datacite.org/.

[6] https://fairsharing.org/.

datasets created and preserved by others, (b) determining if the data submitted by others are usable, and (c) determining if datasets submitted by others are appropriate for reuse. Information professionals can provide training to researchers in areas such as searching data archives, understanding other's metadata, and identifying and using tools for secondary data analysis.

11.1.3   Citation

Data management helps to provide a standardized method for secondary users to cite data and can also be used by data producers to cite their own data as standalone research products. In 2014, a group called **FORCE11**[7] issued a joint declaration that enumerated "eight data citation principles viz. *importance, credit and attribution, evidence, unique identification, access, persistence, specificity and verifiability,* and *interoperability and flexibility* that cover the purpose, function, and attributes of citations and recognize the dual necessity of creating citation practices that are both human and machine understandable" [1]. Furthermore, the Registry of Research Data Repositories[8] helps researchers, funding organizations, libraries, and publishers to check whether a data repository chosen by them supports the creation of unique data citations that embody the joint declaration of data citation principles.

Why Cite Data?

By providing data citation to your work, you make it easier for others to identify and acknowledge your data. Data citation promotes the reproduction of research results, tracks the use and impact of data, and gives a structure that recognizes and rewards the data creator.

Citation Elements

DataCite,[9] an international standard body founded in 2009, works with data repositories to assign persistent identifiers such as digital object identifiers (DOIs) to data. DOIs (i) are effective methods of data citation, discovery, and access, (ii) ensure that data can be discovered online regardless of where they are located, and (iii) provide methods for identification that is machine actionable as well as provide stable access to the data resource using a persistent identifier. It recommends the following minimum citation elements:

- *Creator (Publication Year). Title. Publisher. Identifier.*
- Two additional properties (Version and Resource Type) may also be added: *Creator (Publication Year). Title. Version. Publisher. Resource Type. Identifier.*

---

[7] https://www.force11.org/datacitationprinciples.
[8] https://www.re3data.org/.
[9] https://datacite.org/cite-your-data.html.

11.1.4   Sharing

The primary key players in data sharing are data creators/producers (by making the data available), secondary users, and data repositories (by enhancing the discovery and reuse of data by creating a formal data citation). It helps in transparency, openness, and maximum funders' return on investment. It helps the research community to:

- Reinforce open scientific inquiry
- Support the replication and verification of original results
- Promote new research
- Test alternative or new methods
- Encourage collaboration and multiple perspectives
- Provide necessary teaching resources
- Avoid efforts in duplicate data collection
- Protect against fraudulent or faulty data
- Enhance overall impact and visibility of research projects
- Preserve data for future use

Some of the significant challenges associated with data sharing are:

- It takes time and efforts to make the data shareable.
- Perceived risks from the loss of control of data.
- Data containing sensitive or confidential information need to be de-identified which may affect the secondary usability of secondary analysis.
- Ownership of the data may be unclear or problematic.
- Lack of incentives for sharing data.
- Lack of experience and knowledge in data management.

11.1.5   Need of Data Management for Text Mining

Even though data is generated exponentially, only a small amount of data is known, let alone published and accessible practically. The accessibility and reuse strategies of textual data by humans and machines can differ, especially in the case of text mining. Textual data is generally transformed into structured data with the help of NLP, statistical processing of data, advanced pattern recognition, clustering techniques, machine learning, etc., for further analysis. Therefore, textual data is needed to be provided in the *right* format with the *right* metadata for the machine to understand it and process it further.

Data creation is both a time-consuming and expensive process that includes steps such as data collection and curation, metadata addition, annotation, maintenance, preservation, and legal clearance. In many scientific processes, data can be additionally reused for secondary value services which were not thought of initially when the project was started. Hence, appropriate tools and mechanisms (technical, scientific, legal, social, and organizational) are required to allow efficient access, sharing, reusing, and re-purposing of textual data.

Thus, there are numerous reasons why data management is important, which include:

- Discovery and interpretation of data by other researchers
- Sustaining the value of data by enabling others to verify and build upon the published results
- Facilitating long-term preservation and access to datasets

11.1.6 Benefits of Data Management for Text Mining

1. *For researchers/data provider*: A complete and efficient DMP allows researchers to discover, document, secure, maintain, preserve, and deploy powerful computational facilities; allows interoperability and lawful sharing; provides recognition in the form of ownership, citation, and publicity; provides an added value by allowing reuse and new modes of re-purposing of data; and allows new collaboration. Thus, data management helps researchers to:

- Do better research
- Optimize the use of data during research
- Collaborate with other researchers
- Verify or refine the published results
- Reduce scientific fraud
- Promote the development of new research questions
- Provide resources for training new researchers
- Discourage unintended redundancy
- Ensure that data is preserved for future researchers to discover, interpret, and reuse
- Sustain the value of data

2. *For data users*: Anyone—researchers, private companies, the general public, scientists—can be a data user and can benefit by (i) accessing a substantial amount of data, tools, and technologies; (ii) quickly identifying and accessing the data; (iii) having a permanent persistence record of data in a repository for future use and preservation; (iv) having clear licensing and terms of use; and (v) creating new collaborations.

11.1.7 Ethical and Legal Rules Related to Text Data

Advancements in storage capacity and computing power lead to the exponential growth of textual data, where data collection progressed from paper-based to

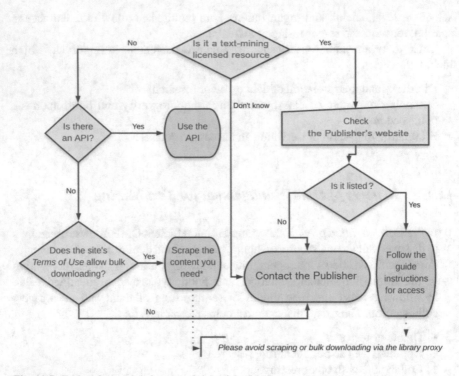

Fig. 11.5 Ethical framework for text mining of databases

digital records. This generation of data leads to new techniques and technologies to transform data to useful knowledge and information. Thus, large databases present some critical ethical challenges that needed to be considered during text mining (Fig. 11.5). Some of the prominent rules to make the data discoverable, interoperable, and reusable not only to humans but also to machines are summarized below.

11.1.7.1   FAIR Data Principles

In order to reuse data, one should comply with FAIR (findable, accessible, interoperable, and reusable) principles, which were first proposed by Wilkinson et al. [3] in 2016 and were adopted by the EU.[10] These principles emphasize the capability of machines to automatically find and use the data, where in this context data is not only data in the conventional sense but algorithms, tools, and workflow. Figure 11.6 summarizes the FAIR principles for research data and should be complied by the researchers and librarians while performing text mining.

---

[10] https://ec.europa.eu/research/participants/docs.

Box 2 | The FAIR Guiding Principles

To be Findable:
F1. (meta)data are assigned a globally unique and persistent identifier
F2. data are described with rich metadata (defined by R1 below)
F3. metadata clearly and explicitly include the identifier of the data it describes
F4. (meta)data are registered or indexed in a searchable resource

To be Accessible:
A1. (meta)data are retrievable by their identifier using a standardized communications protocol
A1.1 the protocol is open, free, and universally implementable
A1.2 the protocol allows for an authentication and authorization procedure, where necessary
A2. metadata are accessible, even when the data are no longer available

To be Interoperable:
I1. (meta)data use a formal, accessible, shared, and broadly applicable language for knowledge representation.
I2. (meta)data use vocabularies that follow FAIR principles
I3. (meta)data include qualified references to other (meta)data

To be Reusable:
R1. meta(data) are richly described with a plurality of accurate and relevant attributes
R1.1. (meta)data are released with a clear and accessible data usage license
R1.2. (meta)data are associated with detailed provenance
R1.3. (meta)data meet domain-relevant community standards

Fig. 11.6 FAIR guiding principles for research data (©2016 Springer Nature, all rights reserved—reprinted under Creative Commons CC BY license, published in Wilkinson et al. [3])

11.1.7.2   Creative Commons

Creative Commons(CC)[11] is a US-based nonprofit organization that is the leader in developing legal tools for sharing creative works over the Internet since the 2000s and is the far most widespread open content licensing model. DataCite[12] recommends using the CC0 license for the data. CC helps to achieve (i) robust legal code, (ii) human-readable summary that is understandable at a glance, (iii) machine-readable layer of code that can help make resources interoperable, and (iv) both copyright and database laws (CC version 4.0). Creative Commons [4] classifies CC licenses into six major categories:

1. **CC BY (Attribution)**: It allows others to distribute, remix, tweak, and build upon your work, even commercially, as long as they credit you for the original work. It is the most accommodating license offered and allows maximum dissemination and use of the licensed work.
2. **CC BY-SA (Attribution-ShareAlike)**: It lets others remix, tweak, and build upon your work even for commercial purposes, as long as they credit you and license their new creations under identical terms. This license is often compared with *copyleft* free and open-source software licenses.
3. **CC BY-ND (Attribution-NoDerivs)**: It allows for redistribution, commercial and non-commercial, as long as it is passed along unchanged and in whole, with credit to you.

---

[11] https://creativecommons.org/.

[12] https://datacite.org/cite-your-data.html.

Additionally, it does not permit adaptations of the work, which may lead to significant problems with the combination of different contents.

4. **CC BY-NC (Attribution-NonCommercial)**: It allows others to remix, tweak, and build upon your work non-commercially. Further, their new works must also be acknowledged by you and be non-commercial but do not have to license the derivative works on the same terms.

5. **CC BY BY-NC-SA (Attribution-NonCommercial-ShareAlike)**: It lets others remix, tweak, and build upon your work non-commercially, as long as they credit you and license their new work under identical terms.

6. **CC BY-NC-ND (Attribution-NonCommercial-NoDerivs)**: It is the most restrictive of all the CC licenses and only allows others to download your work and share them as long as they credit you, but they cannot change the work in any way or use it commercially.

7. **CC0 (No Copyright: Public Domain)**: It lets anyone copy, modify, distribute, and perform the work, even for commercial purposes, all without asking permission.

11.1.7.3   Digital Rights Management

Digital assets are easy to copy, mix, and share and can lead to the dematerialization of the copyrighted materials through digitalization. Finck and Moscon referred *data right management (DRM)* as "software and hardware that defines, protects and manages rules for accessing and using digital content (text, sounds, videos, etc.) by leveraging the exclusive rights recognized by copyright and neighboring rights" [5]. They denoted it "as an additional layer of paracopyright constraints or technological self-protection that annihilate legally recognized protections such as exceptions and limitations in the EU and fair use under US copyright law. It is designed to give maximum control over digital content in self-enforcing conditions and terms of access and use, for instance, publishing and selling of electronic books, digital movies, digital music, interactive games, computer software, and other material in digital format" [5]. Thus, in the context of text mining, DRM follows copyright rules and exceptions (such as fair use that involves search engines, commentary, parody, criticism, research, news reporting, library archiving, teaching, scholarship, etc.) but for digital content specifically.

11.2   Social Media Ethics

It is very difficult to apply standard ethical practices such as *informed consent* and *anonymity* to social media mining. Researchers must develop ethical guidance when using social media data. Some of the critical areas of concern within social media research are:

1. **Is the data public or private?**: As social media platforms post data in public spaces, the users argue that the data is public. Depending upon the platforms, the users (including researchers) often agree to a set of *terms and conditions* stated by third parties. However, many people do not know that the users do not always

read those *terms and conditions*. Therefore, *whether the data is public or private depends upon the context of the data*. For instance, a discussion on Twitter or a post in an open debate can be considered a public data, whereas in the case where people are posting within a private or closed Facebook group, where one has to either become a member or have a password, that data might be considered as a private data.

2. **Do we have to acquire informed consent from the users of social media platforms?**: In most social media platforms, the users are usually not aware that they are being observed by the researchers or their data is being used by them. Therefore, informed consent is not present in the design of research itself as it would have been in traditional research methodology. Again, whether or not to seek informed consent depends upon certain situations, such as if you are working with data that can be considered as private that might also contain some sensitive data (such as users talking about things like illegal activities, marital status, financial status, a disease they are fighting with, etc., that might bring risk to them if that data is exposed to new contexts or new audience), then you need to seek consent before reusing the social media data. However, usually, with very large datasets, it is difficult to get consent from every user, so you have to consider how you would be using the data. Will you just aggregate the data together, analyze it, and present statistical results, or will you be citing individual tweets/units of data? Another important aspect is whether you are working with children or people who are considered as vulnerable adults, in which case again you have to consider the question of informed consent very seriously.

3. **Is there a need to anonymize the data?**: Anonymizing data is again a very problematic ethical concern with social media research than it is with traditional forms of data collection. It again depends on the way you are using the data where you might not give the name of the people when reusing their data in papers/presentations. However, if you use the unit of data presented on the platform in its original wording, then it can quite often be used to trace back that person's profile or to the owner of the social media data. In order to overcome that problem, you need to pay attention to the *terms and conditions* of the platform where you are accessing the data from because all the social media platforms have different sets of *terms and conditions*, and they change regularly as well. For instance, some platforms stipulate that you cannot reword units of data if you are going to use them in your research, and you have to use them word for word that makes it even more difficult to protect anonymity of the users. Moreover, it becomes much more vital to protect the anonymity of the users if those users had an expectation of privacy when they were posting the data or if the data is considered as sensitive or placed them under any risk or harm when exposed in a new context or if you are working with data that was created by children or vulnerable adults.

So, as researchers, it is your responsibility to make sure that when you are using the data of social media users in your research, (i) you do not place those users under any risk of harm, (ii) you make sure that you are protecting your participants from the risk of harm, and (iii) you think about the nature of the data itself and the context from where you took the data. If you are only analyzing the data but not reporting back the data in your output word for word, then that is perhaps less risky. However, if you want to provide quotes and cite the data in its original format, then that is potentially riskier to your participants.

11.2.1   Framework for Ethical Research with Social Media Data

Figure 11.7 summarizes the discussion from Sect. 11.2 and presents a framework in the form of a flowchart to help the researchers to conduct research using social media data ethically. In addition to that, you might also need to consider your role as a researcher and whether you are also participating in the conversation online. Is there a blurring boundary between researchers and participants? Are you taking part in the conversation, and if so, are you doing so openly as a researcher who is seeking data, or are you doing so as another social media user with interest in the discussion taking place? In that case, you really need to think about the ethics of contributing to that conversation and, in some cases, perhaps guide the conversation for your own purposes as a researcher and whether or not to disclose your primary purpose for participating in that group.

Conversation about social media ethics is ongoing because technology is constantly changing, and social media platforms are continuously changing, as are their *terms and conditions*. So, we can hope that new ethical frameworks would emerge and refine with time as technology and its uses change. The framework itself should be considered as guidance but not as prescriptive because each social media context is unique. The ultimate responsibility lies with the researchers and their ethics committee to ensure that the approach they are taking is ethical.

11.3   Ethical and Legal Issues Related to Text Mining

"Law and technology have a complex relationship. Technology shapes legal development, while it is also shaped by law" [5]. Library professionals need to understand the ethical and legal issues related to text mining (Table 11.1). Many LIS job profiles, including university librarian, data librarian, scholarly communication librarian, subject librarian, and research librarian, require the candidate to have a good ethical and legal knowledge of text mining. Even though text mining has enhanced the overall functioning of libraries, there are legal and ethical barriers related to it.

The ethical and legal aspects related to text mining are much more complicated than can be anticipated. Therefore, it is hard to recognize the instances where text

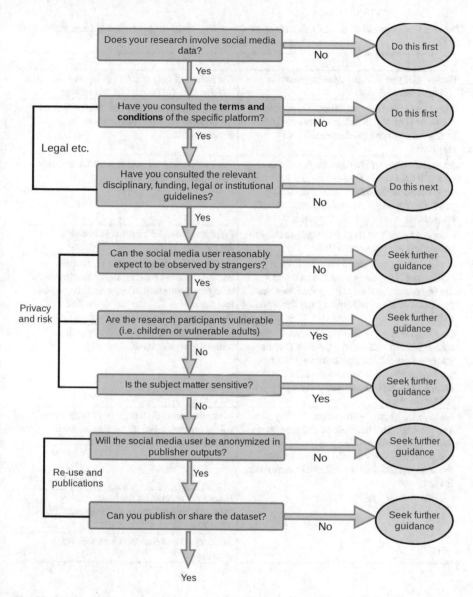

Fig. 11.7 Ethical framework for social media data (adapted from Townsend [6])

Table 11.1 Do's and dont's of using text data for mining by librarians (adapted from McNeice et al. [2])

Do's	Dont's
Establish if you will use or mine data that contains sensitive data	Only think of data protection issues when you usually start the process of text mining
Assign a data protection officer if text mining is one of your organization's core activities, or if your organization does text mining regularly	Collect textual data and just assume it does not concern with personal, private, and sensitive data
Impact assessment (IA): establish what data you will use for what purposes, and who will have access to the data within and outside your organization, and whether your use of textual data brings any legal or ethical risks	Store and retain all data just because it may be helpful in the future
Check whether you have the legal grounds to collect and use the textual data	Randomly transfer or provide access to the data to third parties
Privacy by design: based on your impact assessment (IA), design your whole text mining project in a way that guarantees that you can safely and adequately use the textual data	Reuse textual data from one project to another, without making sure it is compatible with data protection rules, even though you had made sure that the use in the first project is compatible
Look into sector-specific regulation or self-regulation and codes of conduct within your domain, which may provide you more guidance and certainty on what kind of analyses and techniques you can employ on the textual data	Share any textual data with the public, without proper consultation
Anonymize data so you are not dealing with any personal, private, or sensitive data. Note that if you pseudonymize personal, private, or sensitive data, one can still be able to identify the anonymized data if additional information is used	Make decisions affecting the data subjects solely on automated processing of their personal data—it should be prohibited
	Ignore the request by the owner/publisher/funder to access, rectify, or erase data
	Transfer textual data to others without permission

mining becomes unlawful. In the absence of apparent legal exceptions such as fair use, intellectual property rights are most likely to be present for the content created by others. Additionally, it is essential to be cautious in respecting the privacy and personal data of any data subjects. It is always vital to assess the potential legal issues before commencing a text mining project and try to minimize or avoid these issues in your project's design. As all the countries have different laws and regulations related to text mining, this chapter will be restricted to present the text mining rules and regulations for the European Union (EU). Thus, when conducting

Table 11.2 Difference between copyright, neighboring rights, and database rights (©2017, all rights reserved—reprinted under Creative Commons CC-BY license, published in McNeice et al. [2])

Copyright	Neighboring rights	Database rights
Protects the original and creative expressions of the authors	Protects performers (for instance, actors or musicians) and producers of performances or recordings thereof	Protects the investments of producers in creating those databases
It lasts for 70 years after the death of the author (according to EU copyright framework). Historical sources may be out of copyright	It lasts for 50 years after first publications or 70 years in the case of phonograms in the EU	It lasts for 15 years after publication in the EU. If a database is substantially modified, it starts again from the day of the modified version
Examples: Books, websites, research papers, newspaper articles, films, lyrics, musical compositions, original databases, and collections	Examples: Sound recordings, films, broadcasts, fixation of live performance	Examples: Relational databases, SQL databases, tables on a website, playlists on Spotify

text mining research, the two most common legal aspects a data miner should keep in mind are:

1. *Intellectual property rights*, especially copyrights, neighboring rights, and data rights (Table 11.2). It is essential to look at whether the content which you intend to mine is attached to any intellectual property rights or not. If it does, you might need to seek permission for the same from the rights holder. Text mining is most probably to be affected by neighboring rights or copyright, in comparison to data mining as *data* and *facts* are not creative expressions and are free from copyright, except for copyright associated with the collection of the data. Countries like the UK and France let you use the content for text mining without any permission from rights holders but are limited for non-commercial and scientific research purposes. "In many European countries, other text mining exceptions may also exist if you use the content for:

   (a) Private and non-commercial purposes;
   (b) Non-commercial research or teaching purposes in general; and
   (c) Temporary copies are necessary to enable lawful use of work" [2].

2. *Data protection rules*, more specifically the *General Data Protection Regulation (GDPR)*[13] in the EU which replaced the Data Protection Directives in 2018 as a primary law across EU nations to protect personal data. The GDPR rules for data privacy and protection include:

---

[13] https://gdpr-info.eu/.

- The requirement of consent by subjects for data processing
- Protecting the privacy by performing anonymization of the collected data
- Producing notifications for a data breach
- Transferring the data across borders safely
- Appointing a data protection officer to oversee GDPR compliance

The following are some of the vital aspects covered in the EU's data protection rules:

(a) *Personal data*: Personal data is related to an identifiable or identified living person and includes any data that can directly or indirectly let you identify the individual. In the EU specifically, anything from collecting personal data to removing personal data has to comply with European data protection rules. It is essential to get the consent of the data subjects and the consent must be unambiguous, informed, and properly registered. Personal data can be used for research purposes (scientific and historical) in the EU.

(b) *Privacy*: McNeice et al. explained in their FutureTDM report that when "accessing research data made available by other organizations, it is important that mining activities do not inadvertently disclose confidential information or breach the privacy of research subjects. Although the primary responsibility for the ethical collection, storage, and access to research data sits with the research owner, it may be possible to filter data in ways that can reveal confidential or identifying details" [2]. They found that "this is why some data owners require researchers to make an application to use their data or may license its use via a formal agreement or Creative Commons (CC) license" [2].

(c) *Data minimization vs. maximization*: Data maximization is the process of making *big data* text mining so valuable by gathering and using as much data as possible, whereas data minimization allows one to (i) collect personal data only for explicit, specified, and legitimate purposes, (ii) further use the data which is compatible with the purpose it was collected (purpose limitation), and (iii) use only adequate, relevant, and limited data which is necessary for text mining purposes. Hence, as summarized by the library guide of the University of Queensland that "even if the license permits text mining, some approaches to text mining are considered poor etiquette due to the inconvenience they can cause to data providers. For example, bulk scraping or non-rate-limited programmatic querying via APIs can place a significant burden on data providers' servers, causing slow response times or even downtime for other users" [7]. The guide suggested that the "best practice is to check the requirements of the data provider and comply with their preferences regarding data mining activities" [7].

(d) *Sensitive data*: Sensitive data is data related to political opinions, ethnic or racial origin, philosophical or religious beliefs or trade union membership, health data, biometric data, genetic data, and data related to a person's sexual orientation or sex life. European data protection law is stringent for sensitive data and generally prohibits its usage unless you have legal grounds.

(e) *Anonymization of data*: For any text mining project, it is essential to anonymize any personal, sensitive, and private data so that it cannot be reidentified.

11.3.1 Copyright

The library guide of the University of Queensland summarizes how "the process of mining is conducted (e.g., whether the material is copied, reformatted or digitized) without permission could be considered copyright infringement. The ability to data mine relies heavily on technologies that are considered *copy-reliant*, where copies must be made of the data in order for it to be analyzed" [7]. It mentioned that "currently, the Copyright Act 1968 makes no specific exemption for text or data mining. Limited text mining might be covered by the fair dealing exceptions. However, if an entire dataset needed to be copied, this would clearly exceed a *reasonable portion* of the work. While copyright does not apply to raw data or factual information, it does cover the arrangement of data within a database or the *expression* of data" [7].

The Database Directive gives copyright exceptions for personal use, teaching, and scientific research. Thus, the most problematic issue with text mining with library databases, according to Ducato and Strowel, is the "broadness of concept of reproduction (for copyright) and extraction (for the database right). When a text mining tool runs the analysis, it copies all or part of the work and transfers all or substantial part of the content of a database to another medium or adapts or translates the content (such as the conversion of PDF to another format). These operations that are essential in the text mining process fall under copyright or database rights where user cannot perform text mining without the authorization of right holder" [8].

11.3.2 License Conditions

The library guide of the University of Queensland explains how "data providers will have their own specific standards and procedures that you must follow to use the data they provide legally. You must ensure from the outset of your project that the activities you intend to perform during the course of your text mining and the subsequent publication of your research results comply with any licensing *terms and conditions*" [7]. The guide further elucidates how "many data providers license their data to be mined for research purposes only and either prohibit or require special negotiation for text mining with potential commercial applications" [7].

Data miners should ensure that they follow the terms of use of the data while carrying out a text mining project (Table 11.3). Thus, they should look out for different types of Creative Commons categories (refer to Sect. 11.1.7.2) and the restrictions associated with the data in addition to intellectual property rights and data protection rules. Appendix C summarizes text mining licensing conditions for some of the selected prominent publishers.

Table 11.3 Comparison between different license categories

Copyright	Creative commons	Public domain
All rights reserved	Some rights reserved	No rights reserved
NA*		NA*
All *original* work is protected under **Copyright** as designated by the owner	Any work that owners have chosen to be designated as **Creative Commons**	Work published before 1923, work of dead owners, or when owners designate the work as **public domain**
Work *cannot* be adopted, copied, published, or used without the owner's permission	Work *may* be used without permission but as per the rules (above) set by the owner	Work *can* be adopted, copied, published or used without the owner's permission

*NA stands for *not applicable*

11.3.3 Algorithmic Confounding/Biasness

Mass-scale digitization of data, artificial intelligence (AI), and machine learning algorithms led to the automation of simple and complex decision-making processes. Automated data analysis tools, including algorithms used to process text mining, also present ethical challenges. An algorithm can be defined as a set of instructions for how a machine should achieve a particular task. They are the products that involve humans and machine learning. There is long overdue of standards and enforcement of accountability, fairness,[14] and transparency for algorithms in text and data mining procedures. Nowadays, many decisions are governed by algorithms as the datasets are too large to be processed by humans. Thus, humans have become dependent on algorithms to make all kinds of recommendations and decisions for them. They are used to perform the following tasks in the process of text mining and machine learning:

---

[14] Fairness is a nonmathematical, human determination grounded in shared ethical beliefs.

- Ranking information
- Curating information
- Classifying information
- Predicting information
- Clustering information and many other sub-tasks covered in the previous chapters

It may appear that algorithms are neutral and result in unbiased decisions as "they take in objective points of reference and provide a standard outcome" [9]. In reality, we may know what is going inside and coming outside the algorithm, but there is currently no available system of external auditing present that can assess what happens to the data during the process.[15] They are *opinions embedded in mathematics*.[16] As opinions, they are also different where some algorithms such as *word to vector* prefer a specific group of words over another and present incorrect results. "Thus, when an algorithm's output results in unfairness, we can refer to it as bias. Algorithmic accountability is the process of assigning responsibility for harm when algorithmic decision-making results in discriminatory and inequitable outcomes" [9]. Algorithmic accountability helps to determine whether to use a system and whether human judgment is needed to check decisions to counter the biasness (Fig. 11.8).

It is essential to evaluate bias and make the algorithms as accountable as humans making the same decision [11]. In terms of accountability, it is easier to evaluate a non-AI software that applies a simple set of rules, where the process is transparent and can be checked or challenged by those who disagree with the rules, compared to AI software, where most algorithms are opaque and mathematically abstract so that humans cannot understand the rules and cannot challenge them. For instance, a support vector machine (SVM) uses thousand-dimensional space in relation to make thousand-dimensions separators. SVM is understandable in only two or three dimensions but not more. The opaqueness of algorithms can be checked for biasness. Therefore, it is not a problem but can generate unintended bias due to the nature of the data used to run the algorithm. In 2019, Booker [12] proposed the US Algorithmic Accountability Act to force companies to assess the impact of algorithms that make sensitive decisions and correct any found biases. In the same year, the EU released *Ethics Guidelines for Trustworthy AI*[17] outlining seven governance principles: (i) human agency and oversight, (ii) technical robustness and safety, (iii) privacy and data governance, (iv) transparency, (v) diversity, nondiscrimination, and fairness, (vi) societal and environmental well-being, and (vii) accountability. These principles enable inclusion, equal access and treatment, and inclusive design processes in the AI's lifecycle.

---

[15] Pasquale, Frank. 2015. The Black Box Society: The Secret Algorithms That Control Money and Information. Harvard University Press.

[16] O'Neil, Cathy. 2016. Weapons of Math Destruction. Crown.

[17] https://ec.europa.eu/digital-single-market/en/news/ethics-guidelines-trustworthy-ai.

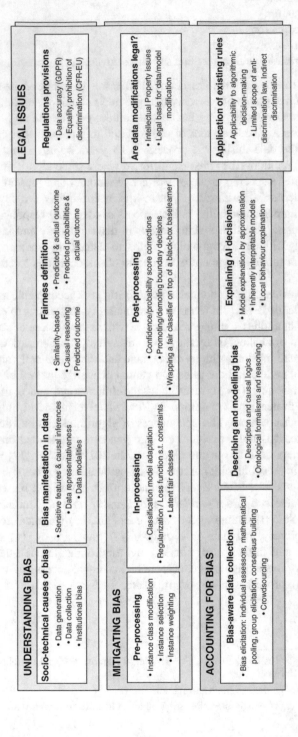

Fig. 11.8 Summary of different issues and solutions related to biasness (©2020 Wiley, all rights reserved—reprinted under Creative Commons CC BY 4.0 license, published in Ntoutseti et al. [10])

There many different ways in which biasness can be created in the algorithm:

- While creating the algorithm by its creator.
- The output of machine learning algorithm often reflects the patterns of input, so if any unexpected results come, it is worth investigating if they are accurate or have inherent bias.
- Insufficient or over-representation of training data.
- The way the algorithm is applied.
- The data on which the algorithm is applied.

11.3.3.1   Types of Biases

1. **Response or Activity Bias**: It occurs in the content generated by humans such as tweets, reviews, posts, etc. As only a small proportion of people within particular demographic groups and geographic areas contribute to opinion, content, and preferences that are unlikely to reflect the opinions of the whole population, this can lead to response or activity bias.
2. **Selection Bias Due to Feedback Loops**: It occurs when a model itself influences the generation of data that is used to train it, such as a ranking system or recommender system. Machine learning models have built-in feedback loops, where the generated data is fed back into the system as training data for the model and is influenced by the user's responses. The user's responses are used to generate label examples by tracking their views, clicks, and scrolls. This bias tends to favor models that generated the data when evaluated using held-out samples from this data and homogenizes user's behavior over time.
3. **Bias Due to System Drift**: It occurs when the model or algorithm changes how users interact with the system over time, for instance, the failure of Google Flu Trends for influenza-like illness based on search data to forecast the expected number of flu cases for a season [13]. It is of two types:

    (a) Concept Drift: It occurs when the target or concept being learned is changed, such as the definition of fraud changes in fraud prediction system.
    (b) Model Drift: It occurs when there is a change in the user interaction model, such as adding new modes of user interaction like share or like buttons or the addition of search assist feature.

    When there is drift in a static model, its performance can degrade over time. Also, if the data from different periods are used to train a model, then the model may not perform well.
4. **Omitted Variable Bias**: It occurs when the vital attributes that influence the outcome are missing and happens when data generation relies on human input, or the process that records the data does not have access to the attributes due to privacy concerns.
5. **Societal Bias**: It occurs when humans produce content on the web and social media, such as gender or race stereotypes.

11.3.3.2 Ways to Mitigate Biases

Though algorithms are deployed to correct humans' biasness, they usually either codify the existing biases or create new ones. They need to be consistently monitored and adjusted with time as they can outdate rapidly. Some of the ways biases can be mitigated are by:

- Understanding how the data was generated
- Performing comprehensive exploratory data analysis (EDA) techniques
- Training the algorithm with sufficient training data which is representative of all the classes
- Using tools that identify bias in models and algorithms such as FairML,[18] IBM AI Fairness 360,[19] Accenture's "Teach and Test" Methodology,[20] Google's What-If Tool,[21] and Microsoft's Fairlearn[22]
- Using R packages for interpretability (such as iml,[23] lime[24,25]) and fairness (such as fairness[26]);
- Making the data, process, and outcome open, thus making it *transparent* and helping us to judge "whether it is fair and whether its outcomes are reliable where different contexts may call for different levels of transparency (EU directive 2016/680 General Data Protection Regulation states the *right to an explanation* about the output of an algorithm)" [9]
- Re-purposing the data and algorithms but keeping in mind the standards that were set and ethical in one setting may create biasness in a new application, therefore creating algorithms and standards that can be adapted from one application to another
- Following the "set of standards proposed by the Association for Computing Machinery US Public Policy Council[27]" [9] and applying them at every stage in the algorithm creation process
- Enforcing accountability in policies during "auditing in pre-and post-processing as well as standardized assessment" [9] as critically algorithms do not make mistakes, but humans do

---

[18] https://dspace.mit.edu/handle/1721.1/108212.

[19] https://aif360.mybluemix.net/.

[20] https://newsroom.accenture.com/news/accenture-launches-new-artificial-intelligence-testing-services.htm.

[21] https://cloud.google.com/ai-platform/prediction/docs/using-what-if-tool.

[22] https://www.microsoft.com/en-us/research/publication/fairlearn-a-toolkit-for-assessing-and-improving-fairness-in-ai/.

[23] https://cran.r-project.org/web/packages/iml/index.html.

[24] https://www.rdocumentation.org/packages/lime/versions/0.5.1.

[25] https://cran.r-project.org/web/packages/lime/lime.pdf.

[26] https://cran.r-project.org/web/packages/fairness/fairness.pdf.

[27] https://www.acm.org/binaries/content/assets/public-policy/2017_usacm_statement_algorithms.pdf.

11.3.3.3 Language-Based Models and Biasness

In recent studies, it has been identified that there are different variety of biases in language-based models (such as sentiment analysis and word embedding) that might disseminate social biases against certain groups of sociodemographic factors (such as race, gender, geography). The output of these language-based models is primarily dependent on the annotated datasets and is sensitive to social bias created by humans. An algorithm that uses both text and metadata to learn is likely to be highly biased as metadata consists of the author's nationality, discipline, etc., when compared to an algorithm with text-only data. Even with text-only data, algorithms will still learn bias due to the language problems generated by second-order effects for text-based machine learning. Additionally, when using chatbots to provide real-time recommendations, the "dialogue of chatbot can be modelled with available metadata to adjust the features of the replier in terms of gender, age, and mood" [14]. Biases in NLP models and tools can emerge from the following sources:

- Amplifying particular views on social media such as "retweeting" the minority opinions and introducing bias
- Creating user bias by putting particular keywords
- Bias emerging from algorithmic decision-making
- Bias by "auto-complete function of search engines" [15]
- "advertisements based on search terms and image search results" [15]
- Biasness created in human-authored data for training the algorithm

Algorithms can magnify and perpetuate the biases if not removed from the confounding factors but can potentially mitigate biases effectively if appropriately designed. Language-based bias can be mitigated by:

- Performing gender-neutral word embedding
- "tagging the data points to preserve the gender of the source" [15]
- Using R packages for interpretability (such as DALEX[28]) and fairness (such as fairness[29])

Appendix C

Text Data and Mining Licensing Conditions

Table C.1 summarizes text mining licensing conditions of some selected prominent publishers.

---

[28] https://cran.r-project.org/web/packages/DALEX/index.html.
[29] https://cran.r-project.org/web/packages/fairness/fairness.pdf.

Table C.1 Selected prominent publishers and their TDM licensing conditions (McNeice et al. (2017) FutureTDM: Reducing Barriers and Increasing Uptake of Text and Data Mining for Research Environments using a Collaborative Knowledge and Open Information Approach. https://project. futuretdm.eu/wp-content/uploads/2017/07/FutureTDM_D5.3-FutureTDM-practitioner-guidelines. pdf. Accessed 5 Nov 2020)

	Elsevier	Wiley	Springer	Emerald
Where is the licence?	http://www.elsevier.com/tdm/userlicense/1.0/	http://olabout.wiley.com/WileyCDA/Section/id-826542.html	http://www.springer.com/tdm	https://www.emeraldinsigh.com/page/tdm
Do I need to check other licenses or documents?	NO The TDM Agreement supersedes any and all prior and contemporaneous agreements	UNCLEAR The click-through TDM agreements say that it supersedes all other prior and contemporaneous agreements, but all that it is superseded by any separate TDM agreement	YES The TDM clause may not have been included in existing SpringerLink subscription agreements but can be added by existing subscribers	NO Not mentioned in policy
Does this license affect my use of open access content?	NO Individual OA licenses supersede anything the contrary in apply, you may use content in accordance with the TDM Agreement	NO If more permissive licenses apply, you may use content in accordance with article-level permissions	NO TDM of OA Springer content is usually allowed without restrictions	NO Not mentioned in policy
Is TDM permitted?	YES	YES	YES	YES
Can I carry out TDM for any purposes?	NO You may not extract, develop or use the dataset for any direct mine or use the dataset for any direct or indirect commercial activity	NO You may only text and data mine Wiley content for non-commercial scholarly research related to specific projects; direct or indirect commercial purposes require prior written consent from Wiley	NO You may only access content for TDM for the purpose of non-commercial research	NO TDM rights are granted purely for internal non-commercial research purposes
Do I need to tell anyone what I am doing with TDM?	MAYBE You must provide TDM output and any related content to Elsevier on request	NO Not mentioned in policy	NO Not mentioned in policy	NO Not mentioned in policy
Are my TDM activities monitored?	YES You are required to use an API key; Elsevier maintains information about you which may be used in aggregate, and may be used to promote Elsevier offers to you	YES You are required to use an API	NO No authentication is required when retrieving SpringerLink content for TDM	NO No authentication is required for CrossRef's TDM API
	Ideal for TDM activities	Close to ideal for TDM activities	Some negative implications for TDM activities	Very restrictive for TDM activities

Note: Table C.1 is colored coded as

	Elsevier	Wiley	Springer	Emerald
Can I access any content I like?	NO — You are licensed to solely to access content made available made available via the CrossRef API	NO — You may only access content via APIs	YES — You may use all subscribed content	YES — Emerald suggests using CrossRef's TDM service to identify and access content
Can I access content any way I like?	NO — You are licensed to use a set of proprietary APIs to access data; you may not use any automated programs to search or may not use any automated programs to search or scrape any Elsevier web site or application	NO — You must access content using a Wiley-approved API; you may not bypass the API; you may not use any automated programs to search or scrape any Wiley content	YES — You are encouraged but not required to download content directly from the SpringerLink platform; friendly DOI-based URLs are provided, tools and methods are suggested, and no API key is required	YES — You are encouraged to use CrossRef's TDM services, but not forbidden from accessing content in other ways
Are there limits on how much content I can access, or how quickly?	UNCLEAR — No details are provided about rate limiting through Elsevier's API	SOMETIMES — You must abide by any rate-limiting which may be conveyed from time to time	VOLUNTARY — You are asked to be considerate and limit your download speed to a reasonable rate	SOME — There are no hard limits on the number of items that may be downloaded, but you may be blocked if your downloading constitutes unfair usage
Do I need to ask or inform anyone before carrying out TDM?	NO — Not mentioned in policy	NO — Not mentioned in policy	NO — Not mentioned in policy	ADVISED — You are advised to inform Emerald you wish to mine their site to avoid being blocked due to unfair usage
Are there limitations on the types of TDM analysis I can perform?	YES — You are licensed to extract semantic entities for the purpose of recognition and classification of relations and classifications between them	NO — You are licensed to carry out computational analysis including but not limited to identification of entities, structures, and relationships	NO — None mentioned in policy	SOME — License includes specific definitions of TDM activities and outputs; these are broad but you must not perform systematic or substantive extracting of content
Are there restrictions on how I can store and share datasets I am using for TDM?	NO — None mentioned in policy	YES — You may load and technically format content on your servers for use for specific TDM projects; you may not otherwise create any form of central repository for Wiley content, or any product or service that could potentially substitute any existing Wiley services	NO — None mentioned in policy	SOME — You may load and technically format XML content on your server, PC or laptop to enable access and use of content for allowed TDM purposes; you may not make results of TDM outputs available on any externally facing server or website

Ideal for TDM activities	Close to ideal for TDM activities	Some negative implications for TDM activities	Very restrictive for TDM activities

Note: Table C.1 is colored coded as

	Elsevier	Wiley	Springer	Emerald
Are there restrictions on how I can share new knowledge I generate as a result of TDM?	YES. Results may be used by you and your company or institution, but may not be used in a way that would compete with existing Elsevier products; a specific proprietary notice must be used when sharing results externally	YES. You may communicate TDM outputs as part of original non-commercial research, including in articles about that research	NO. None mentioned in policy	YES. There are no restrictions on where and how you can publish your research results, but you may not make results of TDM outputs available on any externally facing server or website
Am I required to share the outputs of my TDM research?	YES. You must provide TDM output and any related content to Elsevier on request to ensure compliance with their agreement	NO. None mentioned in policy	NO. None mentioned in policy	NO. None mentioned in policy
Can I support my results with experts from the content I have mined?	YES. Limited to the query-dependent text of a maximum length of 200 characters surrounding the semantic entity matched, or bibliographic metadata; must include a DOI link to the original material	YES. Limited to brief quotations as permitted under national copyright laws; must include a DOI link to the original material	YES. Limited to quotations of up to 200 characters, 20 words, or maximum of 200 characters, one complete sentence; must include a DOI link to the original material	YES. You can use snippets up to a maximum of 200 characters, provided these are referenced as you would reference a copyrighted work; you must contact Emerald if larger extracts are exceptionally required
Can I retain datasets for verifiability and reproducibility of my results?	NO. You may not substantially retain the dataset; all Elsevier content stored for TDM must be permanently deleted on termination of the agreement	NO. You must delete all Wiley content downloaded for TDM on completion of any specific TDM project, or on termination of the agreement with Wiley	UNCLEAR. Not mentioned in policy	NO. You may not substantially retain content; all copies of Emerald content that have been locally loaded for TDM must be destroyed on termination or expiry of this license
Do I have other responsibilities or obligations?	YES. You are responsible for complying with data protection and relevant privacy laws when using or processing personal data	YES. You must implement and maintain data security measures to protect Wiley content in line with international industry standards; you are responsible for complying with data protection and relevant privacy laws when using or processing personal data	NO. None mentioned in policy	NO. None mentioned in policy

| Ideal for TDM activities | Close to ideal for TDM activities | Some negative implications for TDM activities | Very restrictive for TDM activities |

Note: Table C.1 is colored coded as

References

1. Martone M (ed) (2014) Data Citation Synthesis Group: joint declaration of data citation principles, San Diego, CA: FORCE11. https://doi.org/10.25490/a97f-egyk
2. McNeice K, Caspers M, Gavriilidou M (2017) FutureTDM: reducing barriers and increasing uptake of text and data mining for research environments using a collaborative knowledge and open information approach. https://project.futuretdm.eu/wp-content/uploads/2017/07/FutureTDM_D5.3-FutureTDM-practitioner-guidelines.pdf. Accessed 5 Nov 2020
3. Wilkinson MD, Dumontier M, Aalbersberg IjJ, Appleton G, Axton M, Baak A, Blomberg N, Boiten J-W, da Silva Santos LB, Bourne PE, Bouwman J, Brookes AJ, Clark T, Crosas M, Dillo I, Dumon O, Edmunds S, Evelo CT, Finkers R, Gonzalez-Beltran A, Gray AJG, Groth P, Goble C, Grethe JS, Heringa J, 't Hoen PAC, Hooft R, Kuhn T, Kok R, Kok J, Lusher SJ, Martone ME, Mons A, Packer AL, Persson B, Rocca-Serra P, Roos M, van Schaik R, Sansone S-A, Schultes E, Sengstag T, Slater T, Strawn G, Swertz MA, Thompson M, van der Lei J, van Mulligen E, Velterop J, Waagmeester A, Wittenburg P, Wolstencroft K, Zhao J, Mons B (2016) The FAIR Guiding Principles for scientific data management and stewardship. Scientific Data 3:160018. https://doi.org/10.1038/sdata.2016.18
4. Creative Commons (2020) About the licenses. https://creativecommons.org/licenses/. Accessed 6 Nov 2020
5. Finck M, Moscon V (2019) Copyright law on blockchains: between new forms of rights administration and digital rights management 2.0. IIC 50:77–108. https://doi.org/10.1007/s40319-018-00776-8
6. Townsend L (2017) Social media research & ethics. SAGE research methods [streaming video]. SAGE, London. https://doi.org/10.4135/9781526413642. Accessed 26 Feb 2021
7. Berends F (2020) Library guides: text mining & text analysis: considerations - ethics, copyright, licencing, etiquette. https://guides.library.uq.edu.au/research-techniques/text-mining-analysis/considerations. Accessed 6 Nov 2020
8. Ducato R, Strowel A (2019) Limitations to text and data mining and consumer empowerment: making the case for a right to "Machine Legibility." IIC 50:649–684. https://doi.org/10.1007/s40319-019-00833-w
9. Caplan R, Donovan J, Hanson L, Matthews J (2018) Algorithmic accountability: a primer, data & society. https://datasociety.net/wp-content/uploads/2019/09/DandS_Algorithmic_Accountability.pdf. Accessed 8 Nov 2020
10. Ntoutsi E, Fafalios P, Gadiraju U, Iosifidis V, Nejdl W, Vidal M-E, Ruggieri S, Turini F, Papadopoulos S, Krasanakis E, Kompatsiaris I, Kinder-Kurlanda K, Wagner C, Karimi F, Fernandez M, Alani H, Berendt B, Kruegel T, Heinze C, Broelemann K, Kasneci G, Tiropanis T, Staab S (2020) Bias in data-driven artificial intelligence systems—an introductory survey. WIREs Data Min Knowl Discovery 10:e1356. https://doi.org/10.1002/widm.1356
11. Lepri B, Oliver N, Letouzé E, Pentland A, Vinck P (2018) Fair, transparent, and accountable algorithmic decision-making processes. Philos Technol 31:611–627. https://doi.org/10.1007/s13347-017-0279-x
12. Booker C (2019) Booker, Wyden, Clarke introduce bill requiring companies to target bias in corporate algorithms. https://www.booker.senate.gov/news/press/booker-wyden-clarke-introduce-bill-requiring-companies-to-target-bias-in-corporate-algorithms. Accessed 12 Nov 2020
13. Butler D (2013) When Google got flu wrong. Nat News 494:155. https://doi.org/10.1038/494155a
14. Cirillo D, Catuara-Solarz S, Morey C, Guney E, Subirats L, Mellino S, Gigante A, Valencia A, Rementeria MJ, Chadha AS, Mavridis N (2020) Sex and gender differences and biases in artificial intelligence for biomedicine and healthcare. npj Digit Med 3:1–11. https://doi.org/10.1038/s41746-020-0288-5
15. Diaz M, Johnson I, Lazar A et al (2018) Addressing age-related bias in sentiment analysis. In: Proceedings of the 2018 CHI conference on human factors in computing systems. Association for Computing Machinery, New York, NY, pp 1–14

Additional Resources

1. Barocas S, Hardt M, Narayanan A (2019) Fairness and Machine Learning. http://www.fairmlbook.org. Accessed 17 June 2021
2. D'Ignazio C, Klein LF (2020) Data feminism. The MIT Press, United States. https://data-feminism.mitpress.mit.edu/. Accessed 17 June 2021
3. Noble SU (2018) Algorithms of Oppression: How Search Engines Reinforce Racism. New York University Press
4. https://uc-r.github.io/lime
5. https://www.kdnuggets.com/2019/09/python-libraries-interpretable-machine-learning.html
6. https://onthebooks.lib.unc.edu/
7. https://dhdebates.gc.cuny.edu/read/untitled/section/557c453b-4abb-48ce-8c38-a77e24d3f0bd

Index

Printed in the United States
by Baker & Taylor Publisher Services